AF531500

Applications of Remote Sensing and GIS in Geosciences

NIPA® GENX ELECTRONIC RESOURCES & SOLUTIONS P. LTD.
New Delhi-110 034

About the Author

Dr. Vartika Singh is an accomplished researcher and academician with a diverse range of expertise in the field of geology and hydrogeology. Her educational background includes a Ph.D. and M.Phil. degree in Hydrogeology from Vikram University and a Master's degree in Applied Geology from Kurukshetra University. With 15 years of research experience at the Defence Terrain Research Laboratory (DTRL), DRDO, Delhi, and Amity University Noida, she has made significant contributions in her field. Her specialization in research includes Hydrogeology, Remote Sensing, Environmental Geology, Groundwater, Watershed Management, and Artificial Intelligence. Dr. Vartika Singh's extensive research output is commendable. She has published more than 60 research papers, authored three patents, and written three books along with 20 book chapters. Some famous books are, Indian Agriculture- Natural Resource & Sustainability Issues and Benefits of Introducing Soilless Cultivation, Geological vulnerability of Khetri Jodhpur and Nakora Rajasthan India and A Text Book on Ecology and Environmental Sustainability. Her books cover a wide range of topics, from the geological vulnerability of specific regions to broader issues like ecology and environmental sustainability, reflecting her holistic approach to research. This demonstrates her dedication to scholarly work and the wide range of topics she has explored in her research. As an Assistant Professor at AIGWES, Amity University Noida, Dr. Vartika Singh also plays a role in shaping the academic landscape. Her supervision of several research students indicates her commitment to mentoring and guiding the next generation of researchers. Being a part of the editorial boards of several national and international journals reflects her involvement in the peer-review process and her contribution to the academic community's knowledge dissemination. She has impressive research experience, publication record, and active involvement in academia highlight her expertise and significant contributions in the field of hydrogeology, remote sensing, and environmental geology.

Applications of Remote Sensing and GIS in Geosciences

Vartika Singh
Assistant Professor of AIGWES
Amity University
Noida, Uttar Pradesh

NIPA® GENX ELECTRONIC RESOURCES & SOLUTIONS P. LTD.
New Delhi-110 034

NIPA® GENX ELECTRONIC RESOURCES & SOLUTIONS P. LTD.

101,103, Vikas Surya Plaza, CU Block
L.S.C.Market, Pitam Pura, New Delhi-110 034
Ph : +91 11 27341616, 27341717, 27341718
E-mail: newindiapublishingagency@gmail.com
www: www.nipabooks.com

For customer assistance, please contact
Phone: + 91-11-27 34 17 17
Fax: + 91-11-27 34 16 16

Print ISBN: 978-93-58879-71-1

ebook ISBN: 978-93-58874-85-1

Composed and Designed by NIPA®.

Dedicated to

The Sweet Memory of my
Beloved Husband.....

Late Shri Ekank Rana

Who was a source of
Inspiration to me

Preface

Remote Sensing and Geographic Information Systems (GIS) play a crucial role in advancing our understanding of the Earth's geosciences. These technologies provide valuable tools for data acquisition, analysis, visualization, and decision-making in various geoscientific disciplines. The application of remote sensing and GIS in geosciences has revolutionized the way we study and manage natural resources, monitor environmental changes, and assess hazards. Remote sensing involves the collection of data about the Earth's surface and atmosphere using sensors mounted on satellites, aircraft, or drones. These sensors capture images and other measurements in different spectral bands, ranging from visible light to microwave radiation. This data helps scientists study a wide range of geoscience phenomena, including land cover and land use, vegetation dynamics, oceanography, atmospheric conditions, and geological features. GIS, on the other hand, is a powerful tool for capturing, storing, analyzing, and displaying geospatial data. It combines maps, satellite imagery, and various data layers to create a digital representation of the Earth's surface. GIS enables geoscientists to integrate different datasets, perform spatial analysis, and generate models and visualizations that aid in decision-making.

By making the book affordable and user-friendly, you have made it more accessible and valuable to students who are seeking to learn from it.

I appreciate the support and cooperation provided by my family, friends, and colleagues throughout the process of creating this volume. Their sincere assistance has been invaluable. I would like to give a special mention to Shri J.C. Kala, IFS (Retd), Honorary Adviser at the Amity Institute of Global Warming and Ecological Studies, Amity University, Noida. Their guidance and expertise have greatly contributed to the development of this work. I would also like to express my gratitude to Dr. V. K. Panchal, Director (Retd) of DTRL, DRDO, New Delhi, for their contributions and support. Furthermore, I am thankful to Mr. Ekank Rana, Director of Eroth Tech (Mohsri Care Pvt. Ltd.), for their assistance and involvement in the creation of this volume. Their collective efforts and expertise have played a significant role in making this project possible, and I am truly grateful for their contributions.

I am sincerely thankful to NIPA for publishing this book.

Noida
07.03.2024

Vartika Singh

Contents

1

Basics of Remote Sensing and Geosciences

Traditionally, remote sensing encompasses a wide array of technologies and methods used to gather information about objects, areas, or phenomena without direct physical contact. This includes devices like seismographs, fathometers, radar systems, and more. These tools enable us to observe and measure various aspects of the environment from a distance.

However, in contemporary contexts, when we talk about remote sensing in a more specialized sense, we often refer to the use of satellite and aerial-based systems to capture and analyze electromagnetic radiation reflected or emitted from the Earth's surface. This form of remote sensing has become incredibly valuable in fields such as environmental monitoring, urban planning, agriculture, forestry, and disaster management, among others.

Modern remote sensing techniques leverage sensors capable of detecting specific wavelengths of electromagnetic radiation, such as visible light, infrared, and microwave radiation. By analyzing the patterns and intensity of this radiation, researchers can derive valuable information about the Earth's surface features, including vegetation cover, soil moisture, land use, and changes over time.

The term "remote sensing" can encompass a broad range of technologies, in contemporary usage, it's often associated with the advanced methods used to study the Earth's surface from a distance using electromagnetic energy. Remote sensing encompasses various techniques and technologies for acquiring data about Earth's surface and atmosphere without physical contact. It relies on the detection and recording of electromagnetic radiation, which can originate from different sources within the target area or be externally generated and reflected or emitted by the targets.

Remote sensing platforms, such as satellites, aircraft, or drones, are equipped with sensors capable of capturing different wavelengths of electromagnetic radiation. These sensors can detect visible light, infrared, microwave, and other forms of radiation. By analyzing the electromagnetic signatures received by these sensors, remote sensing enables the extraction of valuable information about the Earth's surface features, vegetation, atmosphere, and environmental conditions.

The data obtained through remote sensing can be processed, analyzed, and interpreted to generate maps, monitor changes over time, assess environmental conditions, and support various applications in fields like agriculture, forestry, urban planning, disaster management, and climate studies. Overall, remote sensing plays a crucial role in understanding and managing our planet's resources and environment.

These definitions from the American Society of Photogrammetry and Robbins provide further insight into the diverse techniques and technologies encompassed by remote sensing.

According to the American Society of Photogrammetry, remote sensing imagery is acquired using sensors other than conventional cameras. These sensors capture scenes through electronic scanning and detect radiations outside the normal visual range, including microwave, radar, thermal, infrared, ultraviolet, and multispectral wavelengths. Specialized processing techniques are then applied to interpret this imagery for various applications, including the production of conventional maps.

On the other hand, Robbins, in 1999, defines remote sensing as the non-contact recording of information across different regions of the electromagnetic spectrum, including ultraviolet, visible, infrared, and microwave wavelengths. This recording is done using instruments such as cameras, scanners, lasers, linear arrays, and areal arrays, typically mounted on platforms such as aircraft. The acquired information is then analyzed through visual and digital image processing techniques.

Both definitions highlight the diversity of remote sensing technologies and the range of electromagnetic wavelengths utilized to gather information about the Earth's surface and atmosphere. These techniques play a crucial role in various fields, including environmental monitoring, land management, urban planning, and disaster response. Remote Sensing is the science and art of obtaining information about an object, area or phenomenon through the analysis of data acquired by a device that is not in contact with the object, area, or phenomenon under investigation.

The basis of the remote sensing is the electromagnetic radiation or light which gives the information about the characteristics of a particular object. Each object on Earth's surface interacts with electromagnetic radiation in unique ways, based on its physical characteristics, composition, and structure. This interaction results in a distinct signature or response that can be detected and analyzed by remote sensing instruments.

For instance, different minerals, rocks, soils, vegetation types, and other surface features exhibit specific reflectance or emittance properties across various wavelengths of the electromagnetic spectrum. By measuring and analyzing these signatures, remote sensing techniques can identify and differentiate between different objects or materials on the Earth's surface.

This Principle Forms the Basis for Various Remote Sensing Applications. For example

1. Vegetation Monitoring: Healthy vegetation typically reflects more near-infrared radiation and less visible light compared to unhealthy or non-vegetated areas. This difference in reflectance can be used to monitor vegetation health and density.
2. Geological Mapping: Different rock types have distinct spectral signatures due to variations in their mineral composition. Remote sensing data can be used to map geological formations and identify mineral deposits.
3. Soil Classification: Soils have unique spectral properties based on factors such as moisture content, texture, and organic matter. Remote sensing can help classify soils and monitor changes in soil properties over time.
4. Land Cover Mapping: Remote sensing can differentiate between various land cover types such as forests, water bodies, urban areas, and agricultural fields based on their spectral signatures.

By understanding the unique signatures of different objects and materials, remote sensing enables the extraction of valuable information about the Earth's surface and its processes. This information is crucial for various applications in environmental monitoring, resource management, urban planning, disaster response, and scientific research.

The first Earth-observing satellite in the Landsat program, known as Landsat 1, was launched by the United States on July 23, 1972. Landsat 1 paved the way for continuous and systematic monitoring of the Earth's surface from space. It carried sensors capable of capturing multispectral imagery, which provided valuable data for various applications such as land cover mapping, agriculture monitoring, forestry management, and environmental studies. Following the success of Landsat 1, subsequent missions were launched, each improving upon the capabilities of its predecessors. As remote sensing technology advanced, other countries also developed and launched their own Earth observation satellites for similar purposes. These satellites vary in terms of their sensors, spatial resolution, spectral bands, revisit frequency, and specific applications.

Today, there are numerous Earth observation satellite programs operated by different countries and international organizations, including the European Space Agency's (ESA) Sentinel series, Japan's Advanced Land Observing Satellite (ALOS), China's Gaofen series, India's Resourcesat and Cartosat series, and many others. These satellites contribute to a global network of Earth observation capabilities, providing valuable data for a wide range of scientific, environmental, commercial, and humanitarian purposes. Remote sensing technology in India got impetus following the successful launch of Aryabhata in 1975, Bhaskara 1 (1979) and 2 (1981) satellites incorporating a two –band TV camera system, one in the visible and the other in the near infrared with resolution of about 1km along with a tree frequency passive microwave radiometer system; to carry out remote sensing on an experimental basis. The Indian Remote Sensing programmers began with the successful launching of IRS- IA (1988) and 1RS-1B (1991). Both the satellites are equipped with sensors that acquire multi-spectral data with 72.5m and 36.25m spatial resolution, respectively. The sensing instrument was the Linear Imaging Self –scanning Sensor (LISS). LISS –I acquires the 72.5m data and LISS-II collects the 36.25m data. A second generation of IRS satellite operation started with the launch of IRS-C and IRS-ID in 1995 and 1997, respectively with LISS-III with 23m resolution (70m in the mid-infrared band); a panchromatic sensor, with 5-8 m resolution; and a wide field sensor (WIFS), with 188m resolution. India is one of the leading countries in the world having the capability of launching and manufacturing the satellites. After the mission Chandryan-1, it becomes the fourth country in the world that has the capability to launch missions on other planets. The geostationary satellites are those which have the same rotation period as per the earth and hence stationary with respect to the earth and launched on equator of the earth and launched on equator of the earth at the height of 36000 Km. these satellites are helpful in communication and weather forecasting. As the rotation period is same as per earth, these satellites view the same region of the earth. INSAT series of satellites of India are geostationary satellites. Remote Sensing technology has been potentially explored for retrieving timely reliable spatial information and efficient management of inputs. Ever since satellite remote sensing emerged as potential technology for better managing natural resources.

The stages of Remote Sensing include (Fig.1):

- The Sun is a source of electromagnetic radiation (EMR).

- Energy is transmitted from the source to the earth's surface through the atmosphere.
- EMR interacts with the earth's surface.
- Energy is transmitted from the surface to a remote sensor installed on a platform through the atmosphere.
- The sensor detects energy.
- Transmit sensor data to ground station.
- Processing and analysing sensor data
- Providing final data output for diverse applications.

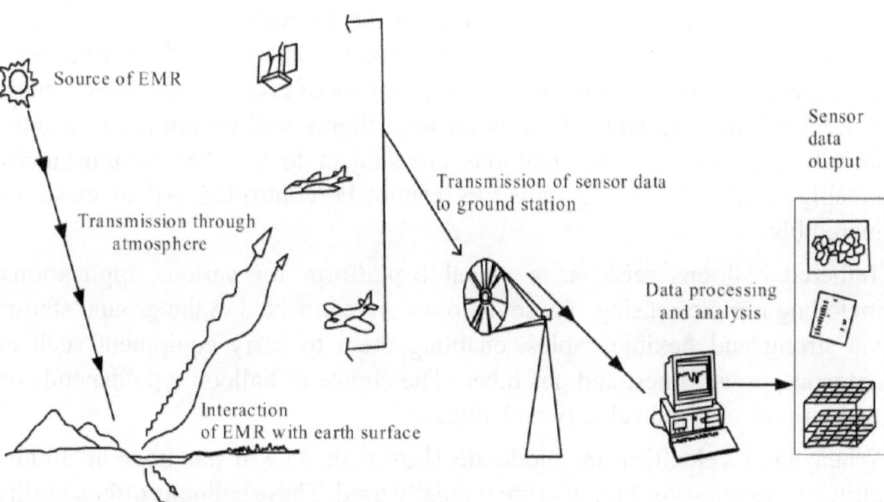

Fig. 1: Stages of Remote Sensing

Scientific Componenets of Remote Sensing

Platforms

Platform refers to the basis on which distant sensors are installed to collect data about the Earth's surface. Platforms can be permanent, such as tripods and balloons, or movable, such as aeroplanes and spaceships. The platforms used are determined by the observing mission's needs and limits. Platforms are classified into three types: 1. Ground borne, 2. Air borne, and 3. Space borne.

1. Ground borne platforms

These platforms are utilised on Earth's surface. The two most typical ground-based platforms are the cherry arm configuration of a remote sensing vehicle and a tripod. They can see the item from various angles and are mostly utilised for gathering ground truth or conducting laboratory simulation investigations.

2. Air borne Platforms

These platforms are placed within the atmosphere of the Earth and can be further classified into balloons and aircrafts.

Balloons

Balloons as platforms are not as costly as aircraft. They have a wide range of forms, sizes, and performance characteristics. The balloons move slowly, do not require any electricity, and vibrate very little. There are three major types of balloon systems: free balloons, tethered balloons, and powered balloons.

Free balloons may reach virtually the top of the atmosphere, providing an intermediate height platform between aeroplanes and spacecraft. Free balloons are capable of lifting thousands of kilogrammes of research payloads. Unless a mobile launching technology is created, flights will be limited to a fixed launching point. The free balloons are subject to weather circumstances, notably winds. The flying trajectory cannot be controlled. All of these are incredibly

Tethered balloons serve as a valuable platform for various applications, including remote sensing. These balloons are connected to the ground station via strong and flexible cables, enabling them to carry equipment such as antennas, power lines, and gas tubes. The choice of balloon type depends on factors such as wind velocity and altitude.

When wind velocities are moderate (less than 35 km per hour at 3000m altitude), sphere-type balloons are typically used. These balloons offer stability and can maintain equipment at a fixed position for extended periods. However, when wind velocities drop further (less than 30 km per hour), natural-shaped balloons become suitable for deployment.

Tethered balloons provide the advantage of keeping equipment in a fixed position for prolonged periods, making them ideal for various remote sensing programs. They can be equipped with sensors and imaging devices to capture data from specific geographic locations over time.

Powered balloons, on the other hand, require propulsion systems to maintain or achieve station over designated areas. These balloons can be remotely controlled and guided along predefined paths or positioned above specific areas within certain limitations. Powered balloons offer greater maneuverability compared to tethered balloons and can be used for tasks such as aerial surveillance, environmental monitoring, and disaster response.

Both tethered and powered balloons play essential roles in remote sensing programs, offering unique capabilities and versatility for gathering data from the Earth's surface and atmosphere.

Aircrafts

Aircraft play a crucial role in remote sensing operations, especially for obtaining aerial photographs and conducting various types of surveys. In India, as you've mentioned, several types of aircraft are utilized for remote sensing missions, each with its own specifications:

Dakota: This aircraft typically operates at a ceiling height of 5.6 to 6.2 km and has a minimum speed of 240 km/hr. It is commonly used for aerial photography and reconnaissance missions.

Avro: With a higher ceiling height of 7.5 km and a minimum speed of 600 km/hr, the Avro aircraft offers enhanced capabilities for remote sensing operations, including aerial surveys and mapping.

Cessna: The Cessna aircraft can reach altitudes of up to 9 km and has a minimum speed of 350 km/hr. It is suitable for various remote sensing applications, such as environmental monitoring and land use mapping.

Canberra: The Canberra aircraft boasts an impressive ceiling height of 40 km and a minimum speed of 560 km/hr. It is utilized for advanced remote sensing tasks, including high-altitude photography and atmospheric research.

In addition to these aircraft used in India, special-purpose aircraft are employed abroad for remote sensing operations, particularly in high-altitude photography:

U-2: This reconnaissance aircraft operates at a ceiling height of 21 km and has a minimum speed of 798 km/hr. It is primarily used for strategic aerial reconnaissance missions, including high-altitude photography and surveillance.

Rockwell X-15: As a research aircraft, the Rockwell X-15 is capable of reaching extreme altitudes, with a ceiling height of 108 kilometers. It achieves incredible speeds of up to 6620 km/hr, making it suitable for conducting scientific experiments and collecting data at high altitudes.

These aircraft, with their diverse capabilities and specifications, play essential roles in remote sensing operations, enabling the acquisition of valuable aerial imagery and data for a wide range of applications.

The benefits of employing aircraft as a remote sensing platform include excellent data resolution, the capacity to carry big payloads, the ability to image huge regions affordably, access to isolated locations, the convenience of picking multiple scales, appropriate control at all times, and so on. However, because to limits in operating altitudes and range, the aircraft is best suited for local or regional programs rather than worldwide surveys. Aside from that, aeroplanes have contributed significantly to the advancement of space-borne remote sensing systems. Sensor testing, as well as the many systems and subsystems that are involved

3. Space borne Platforms

Platforms in space, such as satellites, are not influenced by the earth's atmosphere. These platforms may freely move in their orbits around the Earth. The entire earth, or any section of it, may be covered at regular intervals. The satellite's orbit determines the extent of coverage. We get a great quantity of remote sensing data from these space-borne platforms, and as a result, remote sensing has grown in popularity across the world. Above description provides a clear overview of the two main types of satellite orbits used for remote sensing: geostationary (or Earth synchronous) and sun-synchronous orbits.

1. Geostationary Orbit: Satellites in geostationary orbit are positioned at a fixed point above the Earth's equator, maintaining the same relative position with respect to the Earth's surface. This orbit allows the satellite to appear stationary from the ground observer's perspective. Geostationary satellites are commonly used for weather monitoring, communication, and broadcasting, as they provide continuous coverage of specific regions, such as weather patterns or communication footprints.
2. Sun-Synchronous Orbit: Satellites in sun-synchronous orbit travel from the North Pole to the South Pole while maintaining a consistent angle relative to the Sun. This orbit ensures that the satellite passes over any given point on the Earth's surface at the same local solar time during each orbit. Sun-synchronous orbits are ideal for remote sensing applications because they provide consistent lighting conditions across successive passes, making it easier to compare images taken at different times. These satellites are particularly useful for mapping natural resources, monitoring environmental changes, and conducting disaster monitoring and response activities.

Remote sensing satellites in sun-synchronous orbit capture images of the entire Earth's surface systematically over time, allowing for comprehensive monitoring and analysis. This continuous coverage facilitates various applications, including land use mapping, agriculture monitoring, forest management, urban planning, and disaster risk assessment.

While geostationary satellites offer continuous coverage of specific regions, sun-synchronous satellites provide systematic and consistent global monitoring, making them valuable tools for remote sensing and Earth observation purposes.

Geo-stationary Satellites

Above description of geostationary satellites is accurate and provides a good understanding of their key characteristics and applications. Here's a summary of the points you've mentioned:

1. Orbital Characteristics: Geostationary satellites orbit the Earth above the equator at a fixed height, typically ranging from 36,000 to 41,000 kilometers. They move in the same direction as the Earth's rotation and complete one orbit in approximately 24 hours, synchronized with the Earth's rotation. This synchronization causes the satellite to appear stationary relative to the Earth's surface.
2. Coverage: Geostationary satellites provide continuous coverage of a specific area on the Earth's surface. Their fixed position allows them to maintain constant observation of the same region day and night. However, their coverage is limited to latitudes approximately between 70 degrees north and south, meaning they cannot effectively observe polar regions. Each geostationary satellite can cover approximately one-third of the Earth's surface.
3. Applications: Geostationary satellites are primarily used for communication and weather monitoring purposes. They facilitate telecommunications, broadcasting, and internet services by providing continuous coverage over specific regions. Additionally, they play a crucial role in weather monitoring and forecasting by observing cloud patterns, atmospheric conditions, and weather phenomena in real-time.
4. Examples: Some examples of geostationary satellites include the Indian National Satellite System (INSAT), which serves various communication and meteorological needs in India, as well as the Meteorological Satellite (METSAT) series and the European Remote Sensing (ERS) series.

Geostationary satellites offer valuable capabilities for communication and weather monitoring, providing continuous coverage and real-time data over specific regions of the Earth's surface.

Sun-synchronous Satellites

The description of sun-synchronous satellites provides a clear overview of their key characteristics and applications. Here's a summary based on the points you've mentioned:

1. Orbital Characteristics: Sun-synchronous satellites orbit the Earth from pole to pole, typically at altitudes ranging from 300 to 1000 kilometers. This orbit allows them to pass over any given point on the Earth's surface

at approximately the same local solar time during each orbit. As a result, they maintain a consistent angle relative to the Sun, ensuring uniform lighting conditions across successive passes.

2. Coverage: Sun-synchronous satellites provide global coverage of the Earth's surface on a regular basis. By passing over the same latitude at the same local solar time during each orbit, these satellites offer repetitive coverage of the entire globe. This systematic coverage facilitates various remote sensing applications, including land monitoring, environmental assessment, and disaster management.
3. Applications: Sun-synchronous satellites are widely used for remote sensing purposes, as they offer consistent lighting conditions and regular coverage of the Earth's surface. They provide valuable data for a range of applications, including land use mapping, agriculture monitoring, forestry management, and urban planning. Examples of sun-synchronous satellite missions include the Landsat series, Indian Remote Sensing (IRS) satellites, SPOT series, and NOAA satellites. Additionally, historical missions such as Skylab and the Space Shuttle have also contributed to Earth observation from sun-synchronous orbits.

Sun-synchronous satellites play a crucial role in remote sensing and Earth observation, offering comprehensive coverage and consistent imaging capabilities for monitoring and analyzing various aspects of the Earth's surface and atmosphere.

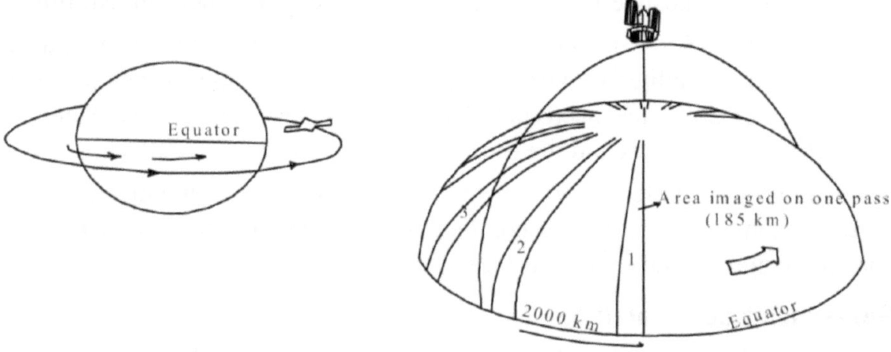

Fig. 2: Geo-stationary and Sun – Syncrhronous Satellites

Remote Sensors

Remote sensors are instruments that detect objects on the earth's surface by measuring the electromagnetic energy reflected or emitted by them. The sensors are installed on the platforms mentioned above. distinct sensors detect distinct wavelength bands of electromagnetic radiation emitted from the earth's surface. As for example, an ordinary camera is the most familiar type of remote sensor which uses visible portion of electromagnetic radiation. Remote sensors can be classed in several ways, as shown below.

On the Basis of Source of Energy Used

Remote sensing sensors can be categorized into two main types based on the source of energy they use: active sensors and passive sensors.

• Active Sensors

Active sensors generate their own energy, which then illuminates the earth's surface. The sensor then receives a portion of this radiation that is reflected back, allowing it to obtain information about the earth's surface. When a photographic camera utilises its flash, it functions as an active sensor. Radar and laser altimeters are active sensors. Radar consists of a transmitter and a receiver. The transmitter sends out a wave that impacts objects and then reflects or echoes back to the receiver.

The characteristics of an active sensor are: 1) Because it produces imagery with both transmitter and receiver units, it consumes a lot of energy. 2) It primarily operates in microwave areas of the EMR spectrum, which may penetrate clouds. 3) It operates in any weather conditions, day and night, and is not dependent on solar radiation. 4) The RADAR signal does not detect colour or temperature information, but it may identify the roughness, slope, and electrical conductivity of the objects being studied.

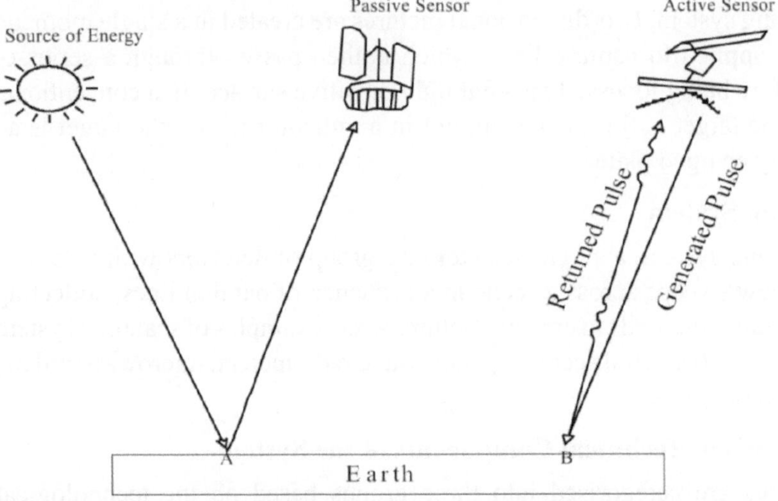

Fig. 3: Active and Passive Sensors

- **Passive Sensors**

Passive sensors do not have their own source of energy and instead rely on external sources of radiation, such as sunlight, to illuminate the target area. The properties of passive sensors include:

1. Simplicity: Passive sensors are typically simpler in terms of both mechanical and electrical design compared to active sensors. They do not require high power requirements since they do not generate their own energy for illumination.
2. Wide Band Systems: Passive sensors often operate across a wide range of wavelengths or wavebands. This wide spectral range allows them to capture a broad spectrum of electromagnetic radiation emitted or reflected by the Earth's surface. This broad coverage helps ensure that useful information is collected even in situations where natural reflectance or emission levels are low.
3. Dependence on Weather Conditions: Passive sensors are dependent on external sources of radiation, primarily sunlight. Therefore, they perform best under good weather conditions when there is sufficient illumination to provide clear images or data.

On the Basis of Function of Sensors

Sensors are classified into two groups based on their function: framing systems and scanning systems.

- **Framing system**

In a framing system, two-dimensional pictures are created in a single moment. A lens is applied to capture light, which is then passed through a series of filters before being focussed on a flat light sensitive surface. In a conventional camera, the target is film emulsion, but in a vidicon camera, the target is an electrically charged plate.

- **Scanning System**

In a scanning system, a single detector or a group of detectors with a defined field of view sweeps across a scene in a sequence of parallel lines, collecting data for continuous cells to create a picture. A few examples of scanning system sensors are multispectral scanners, microwave radiometers, microwave radars, and optical scanners.

On the Basis of Technical Components of the System

The sensors are categorised into three groups based on the technological components of the system and the detecting capabilities. These are: 1.

Multispectral imaging sensor systems, 2. Thermal remote sensing systems, and 3. Microwave radar sensing systems. Multispectral or multiband imaging systems can employ ordinary cameras or a mix of cameras and scanners for different bands of electromagnetic radiation. For example, Landsat's Return Beam Vidicon (RBV) sensor employs both photographic and scanning technologies, comparable to an ordinary TV camera. The thermal system detects temperature changes using radiometers, photometers, spectrometers, and thermometers, whereas microwave sensing systems employ antenna arrays to gather and detect energy from terrain components.

Function of Remote Sensing

A Source of EMR

Electromagnetic radiation (EMR) is crucial for comprehending remote sensing techniques. EMR is a form of energy that travels through space in the form of waves, propagating at the speed of light. These waves consist of electric and magnetic fields oscillating perpendicular to each other and to the direction of propagation. EMR encompasses a broad spectrum of wavelengths, ranging from very short wavelengths like gamma rays to very long wavelengths like radio waves. The classification of EMR types is typically based on their wavelength or frequency. Here are the main types of electromagnetic radiation, organized by wavelength:

1. Gamma Rays: Gamma rays have the shortest wavelengths and highest frequencies in the electromagnetic spectrum. They are emitted during radioactive decay and nuclear reactions, and they possess high energy levels.
2. X-Rays: X-rays have wavelengths shorter than ultraviolet (UV) radiation but longer than gamma rays. They are commonly used in medical imaging and industrial applications for their ability to penetrate materials.
3. Ultraviolet (UV) Radiation: UV radiation has wavelengths shorter than visible light but longer than X-rays. It is emitted by the Sun and can cause sunburn and skin damage. UV radiation is also used in forensic investigations and sterilization processes.
4. Visible Light: Visible light is the portion of the electromagnetic spectrum that is visible to the human eye. It has wavelengths ranging from approximately 400 to 700 nanometers and is responsible for the perception of color.

5. Infrared (IR) Radiation: Infrared radiation has wavelengths longer than visible light but shorter than microwaves. It is emitted by warm objects and is commonly used in night vision devices and thermal imaging.
6. Microwaves: Microwaves have longer wavelengths than infrared radiation but shorter than radio waves. They are used in communication, cooking, and radar systems.
7. Radio Waves: Radio waves have the longest wavelengths and lowest frequencies in the electromagnetic spectrum. They are used in communication systems, such as radio broadcasting, television transmission, and wireless networking.

The characteristics and behaviors of different types of electromagnetic radiation is essential for designing remote sensing systems and interpreting the data collected from them. Different wavelengths of EMR interact differently with the Earth's surface and atmosphere, allowing remote sensing techniques to capture valuable information about the environment.

Kind of waves	**Wavelength Range in micron**
Cosmic Rays	<.0000001
Gamma Rays	.0000001 to .0001
X-rays	.001 to .01
Ultraviolet Light	.01 to .4
Visible Light	.4 to .7
Infra-red Light	.7 to 1000
Microwaves	1000 to 10 6
Radiowaves	more than 10 6

Remote sensing techniques primarily utilize specific portions of the electromagnetic spectrum to gather information about the Earth's surface and atmosphere. The wavelengths most commonly employed in remote sensing applications include:

1. Visible Spectrum (0.4 to 0.7 microns): Visible light, which is the portion of the electromagnetic spectrum visible to the human eye, is frequently used in remote sensing. Sensors capable of detecting visible wavelengths capture information about surface features, vegetation health, and land cover classification. Visible light is particularly useful for producing high-resolution optical images.
2. Reflected Infrared (IR) and Thermal Infrared (0.7 to 14 microns): Remote sensing instruments often utilize both reflected and thermal infrared radiation. Reflected infrared radiation provides valuable data about surface temperature, vegetation health, and moisture content.

Thermal infrared radiation, on the other hand, measures the heat emitted by objects on the Earth's surface. Thermal infrared data are crucial for applications such as land surface temperature mapping, urban heat island analysis, and detecting thermal anomalies.

3. Microwave Region (1000 to 3000 microns): Microwaves, which have longer wavelengths than visible and infrared radiation, are commonly employed in remote sensing for their ability to penetrate through clouds, vegetation, and other atmospheric obstructions. Microwave sensors are utilized for applications such as soil moisture mapping, land surface deformation monitoring (interferometric synthetic aperture radar, InSAR), and sea surface temperature measurements. Additionally, microwave radar systems enable the mapping of terrain elevation and surface roughness.

By harnessing these specific portions of the electromagnetic spectrum, remote sensing technologies provide valuable insights into various Earth processes and phenomena. Each wavelength range offers unique advantages for different types of analyses and applications, contributing to a comprehensive understanding of the Earth's environment.

There are three major sources of EMR utilised for distant sensing that light the item. The Sun is the primary energy source. The sun's radiation spans ultraviolet, visible, infrared, and radio frequency wavelengths. The greatest radiation occurs at.55 microns. All items on the earth's surface emit electromagnetic radiation. The wavelength of these EMR is determined by the temperature and other properties of the item. Many typical items generate EMR with wavelengths ranging from.85 to.95 um. The sensor may also generate EMR at a specified wavelength or region of wavelengths in order to light the item. This is known as active remote sensing, whereas the previous two (EMR from the sun and self-emitted light from the object) are known as passive remote sensing.

Transmission of Energy from the Source to the Surface of the Earth

This passage describes how electromagnetic radiation (EMR) from the sun interacts with the Earth's atmosphere before reaching the surface, affecting image quality in remote sensing. It highlights the significance of certain spectral regions, known as atmospheric windows, where the atmosphere is transparent and EMR can pass through with minimal attenuation. These windows are crucial for satellite-based remote sensing, with commonly used windows (Figure 4) including the visible spectrum (0.38-0.72 microns), near-infrared and middle-infrared (0.72-3.00 microns), and thermal infrared (8.00-14.00 microns) ranges.

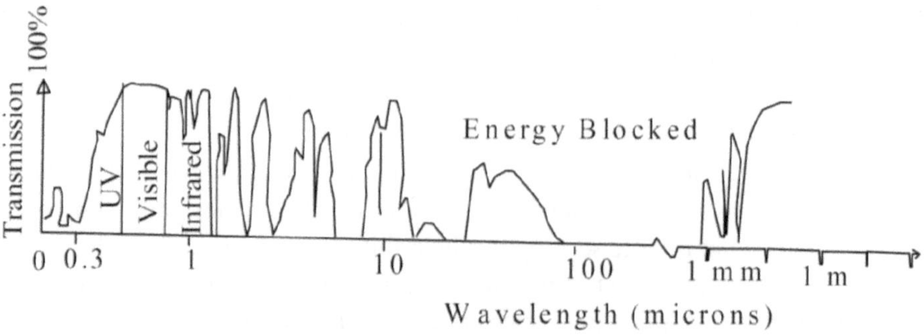

Fig. 4: Atmospheric Windows

Interaction of EMR with Earth's Surface

When electromagnetic radiation interacts with earth, it is either reflected, absorbed, radiated, or transmitted through the substance. It depends on the object's nature and how the wavelength interacts with it. When an object is smoother than the wavelength of the incident radiation, it is reflected in the forward direction, which is known as specular reflection. When the surface is rougher than the wavelength of the incident radiation, it is reflected equally in all directions, a process known as diffused reflection. It should be noted that an item may seem rough in one wavelength area and smooth in another. Fine sand, for example, appears rough in the visible region but smooth in the microwave zone.

Transmission of Energy from Surface to Remote Sensor Mounted on the Platform

EMR travels through the atmosphere after being sent from the earth's surface to the sensor. Here, electromagnetic energy is transformed in the same manner it is transferred from the sun to the earth's surface.

Detection of the Energy by the sensor

Sensors, such as cameras or scanners installed on platforms such as aeroplanes or satellites, recognise and distinguish objects based on the reflected/emitted electromagnetic radiation received by them, known as signatures. The essential notion is that each person has a distinct signature or fingerprint with which he may be identified. Similarly, every item has a unique signal on the sensor that allows it to be identified. In general, a signature is defined as any set of observable qualities that, directly or indirectly, lead to the identification of an item. There are four key features of objects that aid in discrimination: spectral, spatial, temporal, and polarisation changes.

Spectral fluctuations refer to variations in an object's reflectance or emittance as a function of wavelength. The colour of things is a manifestation of spectral diversity in reflectance in the visible area. Vegetation appears green because it preferentially reflects the 'green' component of incoming light. When the sensor employs a wavelength range other than the visible zone, objects do not provide their true colour as a signature. For example, vegetation appears red when the sensor employs near infrared reflectance.

Spatial variations in reflectance and emittance are characteristics of an object's texture, shape, and size. In the visible area of the spectrum, we are well familiar with spatial information offered by an object's shape, size, and so on, as seen in a photograph. Photographs of hot springs will have a comparable form and size to the genuine hot spring. However, the thermal scanner of a hot spring will provide a varied image depending on the temperature distribution throughout the surface of the hot spring.

Temporal fluctuations refer to how reflectivity or emissivity changes over time, such as diurnal and seasonal variations. These changes are significant for detecting agricultural crops and soils. A remote sensor can discriminate between two crops that have identical spectral reflectance but develop during different seasons. In the case of rocks, diurnal fluctuations in emitted radiation (thermal infrared area) have shown to be particularly useful in their identification.

Polarisation varies according to the direction of electromagnetic radiation's electric field component. Objects can be recognised by variations in the polarisation of the reflected EMR. These observations have proven especially valuable in the microwave area.

Transmission and Processing of the Sensor Data Output

When sensors such as cameras or multispectral scanners are installed aboard aeroplanes, the pictures are captured on films or magnetic tapes. When the plane lands, the films or tapes are physically transferred to the laboratory for additional processing. There is no prospect of physically retrieving data-containing media from satellites. In such instance, the data is transferred to a ground station and recorded on magnetic tape. As a result, the ground station must have a suitable tracking antenna and a communication link to the satellite. It undergoes various correction processes to remove distortions and artifacts caused by factors such as motion of the platform (e.g., satellite, aircraft) relative to Earth, platform attitude (orientation), Earth curvature, non-uniformity of illumination, variations in sensor characteristics, atmospheric effects, and more. This is referred to as data processing and can be accomplished via electro-optical methods or, more often, computers. The output goods are then created in either photographic or digital format. The images are then visually analysed, and the digital data is analysed using software. Some common correction methods include:

1. Geometric Correction: This corrects for distortions caused by terrain relief, Earth curvature, and sensor position. Techniques like orthorectification are used to project images onto a standardized map coordinate system.
2. Radiometric Correction: This adjusts for variations in illumination, sensor response, and atmospheric effects. Methods like atmospheric correction remove atmospheric scattering and absorption effects to ensure accurate radiance or reflectance values.
3. Temporal Correction: This aligns data collected at different times to account for changes in environmental conditions, such as lighting and vegetation cover.
4. Sensor Calibration: Regular calibration ensures that sensor readings are consistent and accurate over time. This involves adjusting for sensor biases, drifts, and variations.
5. Motion Correction: This compensates for movement of the platform during data acquisition, ensuring that images are properly aligned and free from blurring or distortion.
6. Noise Reduction: Various techniques are used to reduce noise in the data, such as filtering algorithms and statistical methods.

Remote Sensing Data Products and Their Understanding

Data Products

Social scientists and applied scientists are particularly interested in the data generated by the Remote Sensing technology. Remote sensing data are classified into two types: pictorial and digital. The data products are explained in the following paragraphs:

1. Digital Data Products

Digital data products in remote sensing often take the form of digital images, which are arrays of small cells known as pixels (short for picture elements). Each pixel in the image contains quantitative values representing the electromagnetic energy radiated or reflected from objects within the sensor's field of view. These values can correspond to various properties such as reflectance, brightness, temperature, or other relevant parameters depending on the type of sensor and the application.

Digital images provide a visual representation of the Earth's surface or other observed objects, allowing for detailed analysis and interpretation. They are essential for a wide range of applications, including land cover classification, environmental monitoring, agriculture, urban planning, and disaster management. A digital image is a two-dimensional array of pixels (also known as picture components). Each pixel on the earth's surface represents an area and contains an intensity value (expressed by a digital integer) as well as a location address (linked by its row and column numbers). The intensity value denotes the observed sun radiation in a certain wavelength band reflected from the ground. The location address is a one-to-one correspondence between a pixel's column-row address and the geographical coordinates (such as latitude and longitude) of the imaged place. The digital picture is a large matrix of numbers that is frequently recorded on magnetic tapes, particularly computer-compatible cassettes (CCT). Digital data may be transformed into a photographic picture.

2. Pictorial Data Products

Pictorial data products provide information on items on the earth's surface in the form of pictures or images. Aircraft create graphical goods known as aerial photos. These are often captured by advanced cameras that employ visible portions of electromagnetic spectrum. As a result, aerial images provide an accurate view / image of items on the earth's surface at a reduced size. The aerial images may be in black and white or colour, depending on the camera used in the aeroplane.

Satellite pictures are graphical data products that satellites produce. These pictures are often captured by sensors that employ both visible and invisible portions of electromagnetic radiation. Satellite pictures might be black and white or coloured. Each band of digital data produces a black and white image or picture. A black and white picture for a certain band is created by assigning multiple shades of grey (from white to black) to its digital data. Similarly, unicolor pictures (blue, green, red, etc.) may be created by assigning different shades of blue, green, and red to a certain band data set. When any three bands are joined, it creates multicoloured images. Images taken in blue, green, and red bands (visible portion of electromagnetic). A False Colour Composite (FCC) image is created when photos are captured in the green, red (visible component of electromagnetic energy) and infrared bands (invisible portion of electromagnetic energy) and assigned blue, green, and red colours, respectively, and then blended. The FCC image does not provide a precise picture / view of the earth's surface like aerial images do. The layperson cannot see anything in the FCC graphic. Only a professional can interpret it.

Data Interpretation

In the preceding paragraphs, we discussed two primary forms of Remote Sensing data products: pictorial and digital. Pictorial data products, such as aerial pictures and satellite images, are interpreted visually. Similarly, digital data products or digital pictures are mathematically evaluated by computer software. So, there are two methods to evaluate Remote Sensing data: 1. visually and 2. digitally.

1. Visual Interpretation

Aerial pictures and satellite images are perceived visually. Photogrammetry is the science that studies the interpretation of aerial pictures. To interpret aerial photographs, a variety of sophisticated instruments such as pocket stereoscopes, mirror stereoscopes, and plotters are used in photogrammetry to measure the area, height, and slopes of various parts of the earth photographed, as well as to plot different objects / themes from aerial photographs.

Satellite images are becoming increasingly prevalent as technology advances. Satellite image interpretation involves analysing pictures to identify items and determine their relevance. Interpreters analyse remote sensing images. Use logic to detect, quantify, and assess the relevance of natural and cultural phenomena. Image interpretation needs substantial training and is labor-intensive.

Use logic to detect, quantify, and assess the relevance of natural and cultural phenomena. Image interpretation needs substantial training and is labor-intensive.Size, shape, tone, texture, shadow, pattern, association, and other picture properties are used to extract information from images. Though this approach is simple and straight forward, it has following short comings: i) The range of gray values product on a film or print is limited in comparison to what can be recorded in digital form, ii) Human eye can recognize limited number of colour tones, so full advantage of radiometric resolution cannot be used, iii) Visual interpretation poses serious limitation when we want to combine data from various sources.

2. Digital Interpretation

Digital interpretation enables quantitative examination of digital data using computers to derive information about the earth's surface. Digital interpretation is sometimes known as 'Image Processing'. picture processing includes picture repair, augmentation, and information extraction. The term image correction refers to the process of correcting defects in digital images. Errors occur owing to two factors. When mistakes occur as a result of a sensor defect (for example, if one of 'n' detectors fails to function), this is referred to as a radiometric error. Geometric error occurs when mistakes are caused by factors such as earth rotation, spacecraft velocity, atmospheric attenuation, and so on. Both radiometric and geometric errors/noise in photos are minimised using various computer-aided approaches. Image enhancement is the process of manipulating data to improve its interpretability. Sometimes digital images lack appropriate contrast, making it difficult to recognise distinct things. Thus, the image requires contrast enhancement. Different image enhancement techniques boost contrast in digital images. The ultimate purpose of an interpreter is to extract information from a digital image once it has been corrected, rectified, and contrast enhanced. In Information Extraction, spectral values of pixels are analysed using a computer to detect and classify items on the earth's surface. In other words, spectrally homogeneous pixels in a picture are grouped together and distinguished from others. This allows many earthly properties to be recognised and identified. Field expertise and other kinds of information are also useful in the identification and classification procedures.

Applications of Remote Sensing

The application of satellite imagery has led to the extensive use of imagery by organizations that have an interest in a range of environmental management responsibilities at a state and national level. There are number of applications of remote sensing in different fields. Some of the major applications of remote sensing are in the following areas:

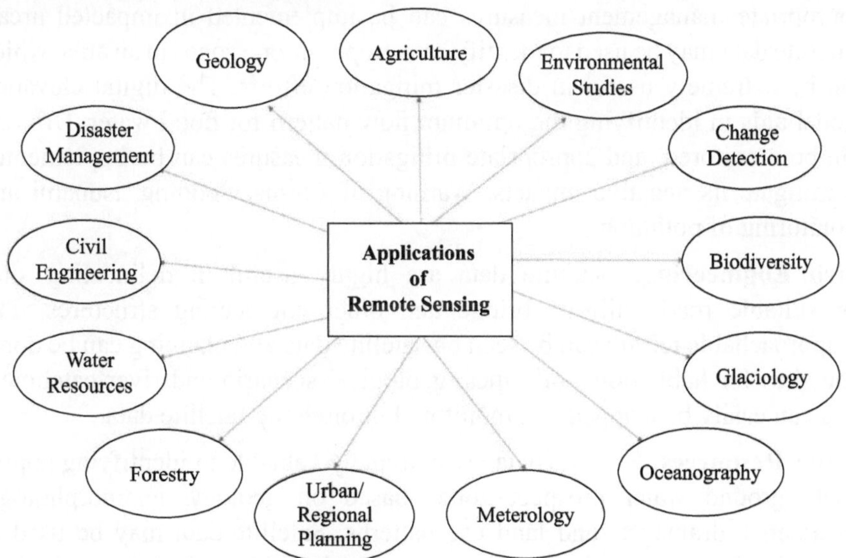

Fig. 5: Different Application of Remote Sensing

Agriculture/ Acreage Forecasting: The satellite data is impartial and provides accurate information about the land characteristics. Repeatability, spectral differentiation, and spatial resolution are some of the useful characteristics of satellite data that aid in acreage forecasting during the rabi and kharif seasons, and these types of studies are extremely useful for estimating production in specific seasons and crops. Regular monitoring of the extend vegetation, soil and there types.

Geology/Mineral Resource: Satellite data are extremely valuable for geological mapping and identifying new mineral resources. New mineral resources can be discovered by mapping their spectral fingerprints. Satellite data may be used to easily map geological locations, particularly those that are difficult to access. Identification of rock types, location of geological faults and anomalies. Its help in mineral and petroleum exploration, mapping of geomorphology and monitoring of active volcanoes.

Environmental Studies: Environmentally degraded lands/regions, water quality concern areas, pollution zones, and point and line pollution sources may all be mapped using satellite data. Satellite data and geographical information systems may be used to study the mapping of dangerous ecological zones, degradation zones, potential environmental consequences, and causes of degradation.

Disaster Management: Disasters may be monitored using satellite data, and appropriate management measures can be implemented in impacted areas. Satellite data may be used to identify catastrophe-prone zones in an area, which can be extremely useful in disaster mitigation efforts. The digital elevation model aids in identifying the optimum flow pattern for flood water. Drought can be monitored, and appropriate mitigation measures can be implemented to mitigate its negative impacts. Warning of storms, flooding, tsunami and monitoring of pollution.

Civil Engineering: Satellite data are highly useful in delineating sites for suitable road, railway, bridge and other engineering structures. The unapproachable terrain can be seen on satellite data and planning can be done. The sites for habitation, soil types, geological scenario and river catchment area can easily be mapped and monitored through the satellite data.

Water Resources: Satellite data are extremely valuable in identifying appropriate ground water prospect zones based on geology, geomorphology, lineaments, drainages, and land use patterns. Satellite data may be used to conveniently manage river basins, watersheds, and surface water bodies. Assessing water resources forecasting melt water run-off from snow.

Forestry: Satellite imagery is used to identify and mapping of species and there native, exotic forest trees, effects of major diseases or adverse change in environmental conditions and geographical extent of forest in area. In unapproachable areas especially in dense and high altitude areas, remote sensing satellite data are highly useful for assessing the forest area, diseases, forest fires, a forestation and deforestation.

Urban/ Regional Planning: Satellite data, geographical information systems, and global positioning systems are extremely valuable in urban/ regional planning. Satellite data provides a comprehensive perspective of the area and aids in the planning of future urban expansion and new habitation locations. Geographical information systems aid in the storage and retrieval of attribute data for a region, allowing for more efficient problem solving. Global positioning systems aid in mapping the exact location of an object, allowing for precise targeting of the desired object.

Meteorology: Profiling of atmospheric temperature and water vapor measuring, wind velocity remote sensing is an effective method for mapping cloud type and extent and cloud top temperature.

Oceanography: Measurements of sea surface temperature mapping ocean currents. Remote Sensing is used for example, to measure sea surface temperature and to monitor marine habitats.

Glaciology: Mapping motion of sea ice and ice sheets determining the navigability of the sea.

Biodiversity: Satellite imagery-derived vegetation type and extent may be integrated with biological and topography data to offer biodiversity information. Typically, this study is performed using a geographic information system.

Change Detection: Satellite imaging cannot usually offer correct information on the species or age of plants. However, the imaging provides an excellent tool of assessing large changes in plant cover, whether caused by removal, wild fine damage, or environmental stress. The most prevalent type of environmental stress is water scarcity.

Land degradation: Imagery can be used to map areas of poor or no vegetation cover. A range of factors, including saline or sodic soils and over grazing can caused degraded landscape.

The field of remote sensing has experienced rapid growth and is increasingly recognized for its importance in recent years. There are two main reasons behind this trend:

1. Growing Interest in Environmental Understanding: There has been a

significant increase in interest among scientists, researchers, students, and the general public in better understanding our environment. People have realized that the geographic space and the events occurring within it are integral parts of our everyday world. Almost every decision we make, whether it's related to urban planning, natural resource management, disaster response, agriculture, or climate change mitigation, is influenced by some aspect of geography. Remote sensing provides a powerful tool for studying and monitoring the Earth's surface and its dynamics, leading to better-informed decisions and sustainable practices.

2. Technological Advancements and Cost Reduction: The advancement of sophisticated space technologies has led to the availability of large volumes of spatial data at increasingly higher resolutions. This includes data from satellites, aerial platforms, drones, and other platforms. Additionally, the costs associated with remote sensing have decreased significantly over time, primarily due to declining costs of computer hardware and software capable of handling and analyzing these vast amounts of data. This makes remote sensing not only accessible to researchers and organizations dealing with complex environmental and spatial issues but also to a wider audience including smaller businesses, government agencies, non-profit organizations, and even individuals.

These two factors combined have contributed to the increasing importance and popularity of remote sensing as a valuable tool for understanding and managing our environment effectively and sustainably.

Remote Sensing in Geosciences

Remote sensing is a powerful tool in the geosciences that involves the acquisition and interpretation of information about the Earth's surface from a distance. This can be accomplished through the use of various sensors mounted on satellites, aircraft, and drones. Remote sensing can provide geoscientists with valuable information about the Earth's surface, atmosphere, and oceans, including data on topography, land use, vegetation, hydrology, geology, and atmospheric conditions. One of the main benefits of remote sensing in the geosciences is the ability to gather data over large areas and in difficult-to-reach locations. For example, satellites can capture data over entire continents, while aircraft and drones can collect data in remote and hazardous environments. This allows geoscientists to study the Earth in a more comprehensive and efficient manner.

Remote sensing data can be analysed using a range of techniques, including image processing, data fusion, and machine learning. This can reveal patterns and relationships in the data that might not be apparent from ground-based measurements alone. Remote sensing can also be used for monitoring and modelling of natural hazards such as earthquakes, volcanic eruptions, and landslides. Remote sensing is a powerful tool in the geosciences that has many applications for understanding the Earth's processes and resources. As the technology continues to advance, it is likely that remote sensing will become an increasingly important part of geoscientific research and applications.

Important Questions

1. What is remote sensing and how is it used in geoscience?
2. What are the different types of remote sensing platforms and sensors used in geoscience applications?
3. How does remote sensing contribute to the study of natural hazards and disaster management?
4. What are some common applications of remote sensing in geology and mineral exploration?
5. How is remote sensing used in agriculture and land management practices?
6. What role does remote sensing play in climate change monitoring and environmental assessment?
7. How can remote sensing techniques assist in mapping and monitoring water resources?
8. What are the advantages and limitations of using remote sensing data compared to traditional field-based data collection?
9. How does remote sensing support urban planning and infrastructure development?
10. How can remote sensing be integrated with Geographic Information System (GIS) for geospatial analysis and decision-making?

2

Lithological Interpretation of Igneous Sedimentary and Metamorphic Rocks

Lithological interpretation of igneous rocks involves the identification and description of the different types of rocks that form as a result of the cooling and solidification of magma or lava. Igneous rocks can be classified based on their mineral composition, texture, and the way they formed. Here are some of the common types of igneous rocks and their characteristics:

1. Basalt

Basalt is a dark-colored, fine-grained rock that forms from the rapid cooling of lava. It is typically composed of minerals such as pyroxene, plagioclase feldspar, and olivine.

Basalt is a common extrusive igneous rock formed from the rapid cooling of basaltic lava. It is typically dark in color, ranging from black to dark gray, and has a fine-grained texture. Basalt is composed mainly of plagioclase and pyroxene minerals, with smaller amounts of olivine, magnetite, and other minerals.

Basalt is found in many parts of the world, including the oceanic crust, volcanic islands, and some continental regions. It is also used as a construction material, such as in road paving, because of its durability and resistance to weathering. Additionally, basalt fiber is used in various applications, including reinforcing materials in composites, insulation, and other industrial uses.

Basalt in India: Basalt is a type of igneous rock that is commonly found in India. It is a dark-colored, fine-grained rock that is rich in iron and magnesium minerals.

In India, basalt is primarily found in the Deccan Plateau region, which covers parts of Maharashtra, Karnataka, Andhra Pradesh, and Tamil Nadu. The Deccan Traps, which is one of the largest volcanic features on Earth, is a vast region in this area that is composed primarily of basalt.

Basalt has been used for a variety of purposes throughout history in India. It is commonly used as a building material for construction of houses, walls, and fortresses, due to its durability and resistance to weathering. Basalt is also used in road construction, as it is a hard and tough material that can withstand heavy traffic.

In addition, the Deccan Traps region has significant geological and scientific importance. The volcanic activity that created the basalt in this region occurred approximately 65 million years ago, coinciding with the extinction of the dinosaurs. The study of the Deccan Traps and its geological history can provide valuable insights into the processes that shaped the Earth's crust and contributed to major events in Earth's history.

Basalt is an important rock in India due to its widespread occurrence and its historical and scientific significance. Its durability and resistance to weathering make it a valuable material for construction and roadbuilding, while its formation provides valuable insights into the geology and history of the region.

Uses: Basalt is a versatile rock with a wide range of uses. Here are some common uses of basalt:

- Construction: Basalt is a durable and strong rock that is commonly used in construction for building walls, roads, bridges, dams, and other infrastructure. It is a popular material for building walls and floors due to its resistance to weathering and erosion.
- Landscaping: Basalt is also used in landscaping and decorative applications. It is commonly used for retaining walls, garden beds, and pathways due to its natural appearance and durability.
- Paving stones: Basalt is a popular material for paving stones due to its durability and skid-resistant surface. It is commonly used in high-traffic areas such as driveways, walkways, and parking lots.
- Filtration media: Basalt is used as a filtration media in water treatment systems due to its high surface area, chemical stability, and ability to remove impurities from water.
- Jewellery: Basalt is used as a gemstone and is often carved into beads, pendants, and other jewellery items due to its attractive black colour and texture.
- Thermal insulation: Basalt fibres are used as thermal insulation material due to their low thermal conductivity, high strength, and resistance to fire.

Basalt is an important rock that has a variety of practical and decorative applications due to its durability, strength, and natural beauty.

2. Andesite

Andesite is an intermediate-colored, fine-grained rock that forms from the cooling of magma. It is typically composed of minerals such as plagioclase feldspar and hornblende.

Andesite is an extrusive igneous rock that forms from the solidification of magma in volcanic environments. It has a fine-grained texture, and its color can range from gray to black, with hints of green, red, or brown. Andesite is typically composed of plagioclase feldspar and one or more mafic minerals such as biotite, hornblende, or pyroxene.

Andesite is named after the Andes Mountains in South America, where it is commonly found. It can also be found in other volcanic regions around the world, such as the Cascades in North America, the Aleutian Islands in Alaska, and the Ring of Fire in the Pacific Ocean.

Andesite is often used as a construction material, especially for decorative purposes such as tiles and facades. It is also used in road construction, as a soil stabilizer, and as a base material for railways. In addition, andesite can be used as a source of mineral nutrients for plants, as it contains potassium, phosphorus, and other elements that are essential for plant growth.

Andesite in India: Andesite is a type of volcanic rock that is composed primarily of plagioclase feldspar and mafic minerals such as pyroxene and amphibole. It typically has a fine-grained texture and is commonly gray, brown, or black in color.

In India, andesite is not as common as other types of volcanic rocks such as basalt. However, it can be found in some parts of the country, particularly in the Western Ghats mountain range along the west coast of India. Andesite is also found in the Himalayan region in northern India, where it is associated with volcanic activity.

Andesite has a variety of uses in India and elsewhere. It is commonly used as a building material for construction of houses, walls, and other infrastructure due to its durability and strength. Andesite can also be used for road construction, as it is a hard and durable material that can withstand heavy traffic.

In addition, andesite has scientific importance in the study of volcanism and plate tectonics. The presence of andesite in certain areas can provide clues about the geological history of the region, including the type and intensity of volcanic activity that occurred.

While andesite is not as common in India as other types of volcanic rocks, it still has practical and scientific applications in construction and geology.

Uses: Andesite is a versatile volcanic rock with a range of practical applications. Here are some common uses of andesite:

- Construction: Andesite is a durable and strong rock that is commonly used in construction for building walls, roads, bridges, and other infrastructure. It is a popular material for building walls and floors due to its resistance to weathering and erosion.
- Decorative stone: Andesite is often used as a decorative stone in landscaping projects due to its attractive appearance and durability. It can be used for garden beds, retaining walls, and pathways.
- Road construction: Andesite is used in road construction due to its hard and durable nature. It can be used as a base material for roads, as well as for paving stones and other road surfaces.
- Stone tools: Andesite has been used for thousands of years to make stone tools, such as knives, scrapers, and arrowheads. It has a fine-grained texture that makes it suitable for precision cutting and shaping.
- Geology: Andesite is an important rock in the study of geology and plate tectonics. It is often associated with volcanic activity and can provide valuable insights into the geological history of a region.

Andesite is an important rock with a variety of practical and scientific applications. Its durability, strength, and attractive appearance make it a popular choice for construction and landscaping projects, while its geological importance makes it valuable for the study of Earth's history and processes.

3. Granite

Granite is a light-colored, coarse-grained rock that forms from the slow cooling of magma. It is typically composed of minerals such as quartz, feldspar, and mica.

Granite is a common intrusive igneous rock that is widely used in construction and monuments. It is composed primarily of quartz, feldspar, and mica minerals, with smaller amounts of other minerals such as amphiboles and pyroxenes. Granite typically has a coarse-grained texture, with visible crystals of its constituent minerals.

Granite forms deep beneath the Earth's surface when magma slowly cools and solidifies. As it cools, the minerals within the magma crystalize and interlock with one another, creating the characteristic speckled appearance of granite.

Granite is known for its durability and resistance to weathering, making it a popular choice for countertops, flooring, and building facades. It is also used in sculptures, monuments, and gravestones. Additionally, granite is used as a source of crushed stone for road construction and other infrastructure projects.

Granite comes in a wide variety of colors and patterns, depending on the specific minerals present in the rock. Some common colors include gray, pink, red, and black. The patterns can range from fine speckles to large swirling patterns, making each piece of granite unique.

Granite in India: Granite is a type of igneous rock that is made up of several minerals, including quartz, feldspar, and mica. It has a hard and durable surface, making it a popular choice for construction and decorative applications. In India, granite is abundant and is found in several states, including Karnataka, Tamil Nadu, and Andhra Pradesh.

Uses: Here are some common uses of granite in India:

- Construction: Granite is commonly used in construction for building walls, floors, and roofs due to its durability and strength. It is often used for interior and exterior flooring, wall cladding, and stairs.
- Monuments and memorials: Granite is a popular material for monuments, gravestones, and memorials due to its durability and ability to withstand weathering and erosion. The famous Taj Mahal monument in Agra, India, for example, features extensive use of white marble and red granite in its construction.
- Countertops: Granite is a popular choice for kitchen and bathroom countertops due to its hard surface and resistance to scratches and stains. It is also available in a wide range of colors and patterns, making it a versatile choice for interior design.
- Landscaping: Granite is often used in landscaping projects, such as for garden beds, retaining walls, and pathways. Its natural appearance and durability make it a popular choice for outdoor use.
- Carvings and sculptures: Granite is a popular material for carving and sculpting due to its hard and durable surface. It can be used to create intricate sculptures, statues, and other decorative objects.

Granite is a versatile rock that has a range of practical and decorative applications in India. Its durability, strength, and natural beauty make it a popular choice for construction, landscaping, and interior design.

4. Diorite

Diorite is an intermediate-colored, coarse-grained rock that forms from the cooling of magma. It is typically composed of minerals such as plagioclase feldspar and hornblende.

Diorite is an intrusive igneous rock that is composed primarily of plagioclase feldspar, biotite, hornblende, and/or pyroxene minerals. It has a coarse-grained texture and is typically gray to dark gray in color.

Diorite is formed when magma slowly cools and solidifies deep beneath the Earth's surface. It is often associated with volcanic arcs and is found in areas of continental crust that have been subjected to tectonic activity.

Diorite is commonly used in construction as a building material and as an ornamental stone. It is used for countertops, flooring, and as a decorative stone for interior and exterior surfaces. It is also used for sculptures, monuments, and gravestones.

Diorite is valued for its strength, durability, and resistance to weathering. It is often used in construction projects where a strong, long-lasting material is needed. It is also used as a source of crushed stone for road construction and other infrastructure projects.

Diorite is similar in appearance to granite, but is typically darker and contains more hornblende and biotite minerals. It can also contain small amounts of metallic minerals such as magnetite or ilmenite.

Diorite in India: Diorite is a coarse-grained igneous rock that typically contains a mixture of plagioclase feldspar, biotite, hornblende, and other minerals. While diorite can be found in many locations around the world, I do not have specific information about diorite found in India.

India is a large and geologically diverse country with a variety of rock formations and mineral resources. Some common types of rocks found in India include granite, basalt, limestone, sandstone, and shale. Diorite may be found in certain regions of India, particularly in areas with volcanic activity or where there are intrusive igneous rocks.

If you have more specific information about the location or context in which diorite was found in India, I may be able to provide more detailed information.

Use: Diorite has been used for a variety of purposes throughout history, including as a building material and as a decorative stone. Its durability, hardness, and resistance to erosion make it a popular choice for construction projects, particularly in areas where weathering and erosion are a concern.

- In ancient times, diorite was used extensively in the construction of monumental structures such as temples, pyramids, and other public buildings. The Egyptians, for example, used diorite to make statues and other ornamental objects. The Olmec civilization of Mesoamerica also used diorite to carve colossal heads that are now some of the most famous artifacts of their culture.

- Today, diorite is still used in construction and landscaping projects, particularly as a crushed stone or aggregate for concrete and asphalt. It is also used as a decorative stone in interior and exterior design, such as for countertops, floor tiles, and wall cladding.

In addition to its practical uses, diorite is also valued by collectors and enthusiasts of mineral specimens and lapidary arts for its interesting patterns, colors, and textures.

5. Gabbro

Gabbro is a dark-colored, coarse-grained rock that forms from the slow cooling of magma. It is typically composed of minerals such as plagioclase feldspar and pyroxene.

Gabbro is a coarse-grained, dark-colored intrusive igneous rock that is composed primarily of calcium-rich plagioclase feldspar, pyroxene, and sometimes olivine. It is typically dark green to black in color, with a phaneritic texture (meaning its crystals are large enough to be seen without a microscope).

Gabbro is formed when magma slowly cools and solidifies deep beneath the Earth's surface. It is often associated with oceanic crust, and is found in areas where tectonic plates are moving apart, such as mid-ocean ridges.

Gabbro is used in construction as a building material and as a decorative stone. It is commonly used for countertops, flooring, and as a facing stone for interior and exterior surfaces. It is also used for monuments, gravestones, and sculptures.

Gabbro is valued for its strength, durability, and resistance to weathering. It is often used in construction projects where a strong, long-lasting material is needed, such as in road construction and seawalls. It is also used as a source of crushed stone for concrete and asphalt production.

Gabbro is similar in composition to basalt, but has a much coarser texture due to its slow cooling rate. It is often found in the same areas as basalt, but is typically deeper beneath the Earth's surface.

Gabbro in India: Gabbro is a coarse-grained, dark-colored intrusive igneous rock that is composed mainly of pyroxene and plagioclase feldspar. It is commonly found in oceanic crust and some continental areas.

In India, gabbro is present in various parts of the country, particularly in areas with volcanic and intrusive activity. For example, the Deccan Traps in western India is a large volcanic province that is predominantly composed of basalt and gabbroic rocks. The Aravalli Range in northwestern India is another region where gabbroic rocks are found.

Gabbroic rocks in India are mainly used for construction purposes, such as for road construction and building materials. They are also used as decorative stones and for landscaping projects. In addition, gabbro is sometimes used as a source of crushed stone for concrete and asphalt production.

The use of gabbro in India is relatively limited compared to other rocks and minerals found in the country. However, it remains an important resource for certain industries and applications.

Use: Gabbro is a versatile rock with a range of uses in various industries. Here are some common uses of gabbro:

- Construction: Gabbro's hardness, durability, and resistance to weathering and erosion make it an ideal material for construction. Crushed gabbro is commonly used as an aggregate in concrete and road construction.
- Dimension stone: Gabbro is a popular choice for decorative and architectural purposes due to its unique texture and color. It is commonly used as a decorative stone for countertops, flooring, and cladding in buildings.
- Monumental stone: Gabbro has been used for centuries for monumental purposes, such as for sculptures and gravestones. Its dark color and durability make it a popular choice for these applications.
- Mineral specimens: Gabbro is a popular rock among collectors and enthusiasts of mineral specimens due to its interesting patterns and textures.
- Industrial applications: Gabbro is used in various industrial applications such as abrasives, refractories, and insulation.
- Jewelry: Gabbro is sometimes used as a gemstone due to its unique texture and color.

Gabbro is a widely used rock with a range of practical and decorative applications in various industries.

Lithological interpretation of igneous rocks can be aided by the use of petrographic microscope to examine thin sections of the rocks. In addition to visual inspection, various analytical techniques such as X-ray diffraction and electron microprobe analysis can be used to identify the minerals present in the rocks, which can further aid in the interpretation of the lithology. Understanding the lithology of igneous rocks is important in geology, as it provides insights into the Earth's geologic history, tectonic activity, and volcanic hazards.

Lithological Interpretation of Sedimentary Rocks

Lithological interpretation of sedimentary rocks involves the identification and description of the different types of rocks that form as a result of the accumulation and lithification of sediment. Sedimentary rocks can be classified based on their texture, grain size, and mineral composition. Here are some of the common types of sedimentary rocks and their characteristics:

1. Sandstone

Sandstone is a clastic sedimentary rock that is composed of sand-sized grains. It is typically well-sorted and can be further classified based on the mineralogy of the grains.

Sandstone is a sedimentary rock composed mainly of sand-sized mineral particles or rock fragments. It is formed by the accumulation and consolidation of sand grains that have been transported and deposited by wind, water, or ice. The sand grains are typically composed of quartz, feldspar, mica, and other minerals, and are held together by a natural cementing material, such as silica, calcium carbonate, or iron oxide.

Sandstone is a versatile rock that is commonly used in construction, as it is durable, easy to work with, and aesthetically pleasing. It can be found in a range of colors, including white, tan, yellow, red, brown, and gray, and can be cut and shaped into a variety of forms, such as blocks, tiles, and slabs. Some famous examples of sandstone architecture include the Petra ruins in Jordan and the Red Fort in Delhi, India. Sandstone can also be used as a decorative stone in landscaping, as well as in the production of glass, ceramics, and abrasives.

Sandstone in India: Sandstone is a sedimentary rock composed of sand-sized grains of mineral, rock, or organic material that have been compacted and cemented together. Sandstone is a common rock type found in many parts of India, particularly in regions with extensive sedimentary deposits.

Sandstone in India: Sandstone is mainly used for construction purposes. Its durability, ease of cutting and carving, and wide range of colors make it a popular choice for building and ornamental purposes. It is commonly used as a building material for walls, columns, and flooring. Some of the major sandstone producing states in India are Rajasthan, Madhya Pradesh, and Uttar Pradesh.

In addition to its use in construction, sandstone is also used for decorative purposes. It is commonly used for garden ornaments, statues, and fountains. Its unique textures and colors make it a popular choice for interior and exterior design.

Sandstone is also used in the production of glass, ceramics, and other industrial products. It is a source of silica, which is used in the production of glass, as well as a source of raw material for the production of cement.

Sandstone is a versatile rock with a range of practical and decorative uses in various industries in India.

Uses: Sandstone is a versatile rock with a range of practical and decorative uses in various industries. Here are some common uses of sandstone:

- Construction: Sandstone is a popular choice for building and construction due to its durability, strength, and natural beauty. It is commonly used as a building material for walls, columns, and flooring.
- Landscaping: Sandstone is a popular choice for landscaping projects such as garden paths, patios, and retaining walls due to its unique textures and colors.
- Decorative purposes: Sandstone is commonly used for decorative purposes such as garden ornaments, statues, and fountains. Its unique textures and colors make it a popular choice for interior and exterior design.
- Glass production: Sandstone is a source of silica, which is used in the production of glass.
- Ceramics and pottery: Sandstone is used as a raw material for the production of ceramics and pottery due to its high content of clay minerals.
- Industrial uses: Sandstone is also used in the production of cement and other industrial products.

Sandstone is a widely used rock with a range of practical and decorative applications in various industries.

2. Shale

Shale is a fine-grained sedimentary rock that is composed of clay-sized particles. It is typically characterized by a fissile texture, meaning that it can easily split along bedding planes.

Shale is a fine-grained sedimentary rock composed mainly of clay minerals such as illite, kaolinite, and smectite, as well as small amounts of other minerals such as quartz, feldspar, calcite, and pyrite. It is formed from the accumulation and compaction of clay, silt, and organic matter on the ocean floor or in freshwater lakes and swamps.

Shale is characterized by its ability to split into thin, flat layers, or "shales," which are often used as roofing tiles or as a building material. Shale also has important industrial applications, including as a source of natural gas and oil, as well as a material for manufacturing cement and bricks.

Shale is an important source of hydrocarbons, as it contains significant amounts of organic matter that can be transformed into oil and gas through the process of thermal maturation. Shale oil and gas are extracted through hydraulic fracturing, or "fracking," which involves injecting high-pressure fluids into the rock to release the hydrocarbons.

Shale can also have a significant impact on the environment, as the process of fracking has been associated with water contamination and seismic activity. However, shale remains an important resource for energy production and industrial applications, and efforts are being made to develop more sustainable and environmentally friendly methods of extracting these resources from shale formations.

Shale in India: Shale is a fine-grained sedimentary rock that is composed of clay minerals, silt-sized particles, and other minerals. Shale is found in various parts of India, particularly in regions with extensive sedimentary deposits.

In India, shale is mainly used for its energy content as a source of natural gas and oil. Shale is a source rock for hydrocarbons and is commonly found in sedimentary basins in India, particularly in the western part of the country. The development of hydraulic fracturing techniques has led to the increased extraction of shale gas and oil in India.

In addition to its use as a source of hydrocarbons, shale is also used in construction as a raw material to produce bricks, tiles, and other building materials. Shale has a high clay content, which makes it suitable for brickmaking.

Shale is also used for environmental remediation purposes. It can be used to cap and seal landfills and contaminated soil to prevent the spread of contaminants.

Shale is a rock with a range of practical applications in various industries in India, particularly in the energy and construction sectors.

Uses: Shale is a versatile rock with a range of uses in various industries. Here are some common uses of shale:

- Energy production: Shale is a source rock for hydrocarbons and is used as a source of natural gas and oil. Advances in hydraulic fracturing techniques have made it easier to extract shale gas and oil, making it an important resource for energy production.
- Construction: Shale can be used as a raw material to produce bricks, tiles, and other building materials. Its high clay content makes it suitable for brickmaking, and it can also be used as a raw material for the production of cement.

- Environmental remediation: Shale can be used to cap and seal landfills and contaminated soil to prevent the spread of contaminants. Its impermeable nature makes it useful for preventing the spread of pollutants and protecting groundwater.
- Art and crafts: Shale's unique texture and color make it a popular choice for art and crafts. It can be carved, polished, and used to make decorative objects, sculptures, and jewelry.
- Research: Shale is used in scientific research as a reference material for studying the properties and behavior of sedimentary rocks.

Shale is a widely used rock with a range of practical and decorative applications in various industries.

3. Limestone

Limestone is a chemical sedimentary rock that is composed of calcium carbonate. It is typically formed from the accumulation of shell fragments or from the precipitation of calcium carbonate from solution.

Limestone is a sedimentary rock composed mainly of calcium carbonate (CaCO3), usually in the form of calcite or aragonite. It is formed from the accumulation of the remains of marine organisms such as coral and shellfish, as well as from the precipitation of calcium carbonate from water.

Limestone is typically light-coloured, but can also come in a range of colors, including white, gray, yellow, and pink. It is a versatile rock that is used in a variety of applications, including construction, agriculture, and industry.

In construction, limestone is commonly used as a building material for walls, flooring, and decorative elements, due to its durability and aesthetic appeal. It is also used as a material for producing cement, concrete, and asphalt, as well as for creating sculptures and other works of art.

In agriculture, limestone is used to neutralize soil acidity and provide essential nutrients such as calcium and magnesium. In industry, limestone is used as a raw material in the production of steel, glass, and other materials, and is also used to purify water and treat industrial wastewater.

Limestone is also an important geological resource for understanding Earth's history and the evolution of life, as many fossils and ancient formations are found within limestone deposits.

Limestone in India: Limestone is a sedimentary rock composed mainly of calcium carbonate, usually in the form of calcite or aragonite. It is a common rock type found in many parts of India, particularly in regions with extensive sedimentary deposits.

In India, limestone is mainly used for construction purposes. Its durability, versatility, and affordability make it a popular choice for building and construction. It is commonly used as a building material for walls, floors, and roofs, as well as for decorative purposes. Some of the major limestone producing states in India are Madhya Pradesh, Rajasthan, and Andhra Pradesh.

In addition to its use in construction, limestone is also used in various industrial applications. It is used as a raw material for the production of cement, steel, and other industrial products. It is also used in the production of lime, which is used in the construction industry, as well as for soil stabilization and water treatment.

Limestone is also used for agricultural purposes, particularly for soil amendment. It is used as a source of calcium and magnesium, which are essential nutrients for plant growth.

Limestone is a versatile rock with a range of practical and industrial uses in various industries in India.

Uses: Limestone is a versatile rock with a range of practical and industrial uses. Here are some common uses of limestone:

- Construction: Limestone is a popular building material due to its durability, versatility, and affordability. It is commonly used as a building material for walls, floors, and roofs, as well as for decorative purposes.
- Industry: Limestone is a major raw material for the production of cement, steel, and other industrial products. It is also used as a flux in the production of iron and steel, and as a raw material for the production of lime.
- Agriculture: Limestone is used for soil amendment and as a source of calcium and magnesium, which are essential nutrients for plant growth. It is used to neutralize soil acidity and to improve soil structure.
- Water treatment: Limestone is used in water treatment to remove impurities and to adjust the pH of water. It is also used in the treatment of acid mine drainage.
- Mining: Limestone is used as a raw material in the mining industry for the extraction of metals such as copper, gold, and zinc.
- Environmental remediation: Limestone is used for environmental remediation purposes, such as the treatment of contaminated soil and the neutralization of acid mine drainage.

Limestone is a widely used rock with a range of practical and industrial applications in various industries. Its versatility and abundance make it an important resource for many sectors of the economy.

4. Conglomerate

Conglomerate is a clastic sedimentary rock that is composed of rounded pebbles or cobbles. It is typically poorly-sorted and can be further classified based on the size of the particles.

Conglomerate is a sedimentary rock composed of rounded, coarse-grained rock fragments that are greater than 2 millimeters in diameter. The rock fragments are typically cemented together by a finer-grained matrix, which can be composed of minerals such as quartz, calcite, or clay.

Conglomerates are formed by the deposition and subsequent consolidation of sedimentary material, such as gravel, sand, and mud, that has been transported by water or gravity. The rounded rock fragments in conglomerate are typically derived from pre-existing rock formations, and can include a wide variety of rock types, such as granite, quartzite, sandstone, and limestone.

Conglomerate can have a range of colors, depending on the composition of its rock fragments and cementing material, and can be found in a variety of geological settings, such as river channels, alluvial fans, and beach deposits.

In addition to its use as a decorative stone, conglomerate has a number of industrial applications. It is sometimes used as a source of gravel for construction and road building, and can also be used as a raw material in the production of concrete and cement. Conglomerate can also be used as a reservoir rock for oil and gas, as the rock's porosity and permeability can allow oil and gas to accumulate and be extracted.

Conglomerate in India: Conglomerate is a sedimentary rock composed of rounded fragments of rocks and minerals that are larger than two millimeters in diameter. It is commonly found in many parts of India, particularly in regions with extensive sedimentary deposits.

In India, conglomerate is mainly used as a decorative stone for landscaping and construction purposes. Its unique texture and color make it a popular choice for garden paths, retaining walls, and other landscaping features. It is also used as a building material for walls and other structures, particularly in regions where it is abundant.

In addition to its use in construction and landscaping, conglomerate is also used in the production of aggregate for concrete and road construction. It is also used as a source of gravel and sand for construction purposes.

Conglomerate is a rock with a range of practical and decorative applications in various industries in India. Its unique texture and color make it a popular choice for landscaping and construction, and its abundance makes it an important resource for the construction industry.

Uses: Conglomerate is a sedimentary rock composed of rounded fragments of rocks and minerals that are larger than two millimeters in diameter. Here are some common uses of conglomerate:

- Construction: Conglomerate is used as a building material for walls and other structures. Its durability and resistance to weathering make it a popular choice for construction in regions where it is abundant.
- Landscaping: Conglomerate's unique texture and color make it a popular choice for garden paths, retaining walls, and other landscaping features. It is also used for decorative purposes in gardens and parks.
- Road construction: Conglomerate is used as a source of aggregate for concrete and road construction. Its hardness and durability make it suitable for use in road construction, where it can withstand heavy traffic and weathering.
- Industrial applications: Conglomerate is sometimes used as a source of gravel and sand for industrial applications such as concrete production, asphalt production, and glass manufacturing.
- Geological research: Conglomerate is studied by geologists to gain insight into the history of the earth's surface. The size and composition of the rock fragments can provide clues to the processes that formed the rock.

Conglomerate is a rock with a range of practical and industrial uses in various industries. Its unique texture and durability make it a popular choice for construction and landscaping, and its use as a source of aggregate makes it an important resource for the construction industry.

5. Coal

Coal is an organic sedimentary rock that is formed from the accumulation and lithification of plant remains. It is typically characterized by a black, combustible texture.

Coal is a fossil fuel that is formed from the remains of ancient plants and animals that lived millions of years ago. Over time, these organic materials were buried and subjected to high pressure and temperature, which caused them to undergo chemical and physical changes and transform into coal.

Coal is composed mainly of carbon, along with small amounts of other elements such as hydrogen, sulfur, oxygen, and nitrogen. It is a black or brownish-black rock that is usually found in layers, or seams, in underground mines or in surface deposits.

Coal is a major source of energy, and is used primarily for electricity generation, as well as for industrial processes such as steel production, cement manufacturing, and chemical production. It is also used as a fuel for heating and cooking in households and businesses.

Coal can be classified into four main types based on its carbon content and heating value: lignite, sub-bituminous, bituminous, and anthracite. Anthracite, which has the highest carbon content and heating value, is the most valuable and is used primarily in industrial applications, while lignite, which has the lowest carbon content and heating value, is the least valuable and is used primarily for electricity generation.

Coal mining can have significant environmental impacts, including land degradation, water pollution, and greenhouse gas emissions. Efforts are being made to develop cleaner and more sustainable methods of energy production and reduce the environmental impacts of coal mining and combustion.

Coal in India: India has a significant amount of coal reserves, making it the third-largest coal producer in the world after China and the United States. Coal is an important natural resource in India, and it is used for a wide range of industrial and domestic purposes.

Uses: Here are some of the main uses of coal in India:

- Power generation: Coal is the primary source of energy for power generation in India. The majority of electricity in the country is generated by coal-fired power plants. Coal is also used in the production of thermal energy for industrial processes.
- Steel production: Coal is used as a primary raw material in the production of steel. It is used as a reducing agent to remove oxygen and other impurities from iron ore during the steelmaking process.
- Cement production: Coal is used as a fuel in the production of cement. It is burned at high temperatures to heat the kiln, which is used to produce cement clinker.
- Domestic fuel: Coal is used as a domestic fuel for cooking and heating in many parts of India. In rural areas where electricity is not available, coal is often used as the primary source of fuel for households.
- Fertilizer production: Coal is used as a raw material in the production of fertilizer. It is used to produce ammonia, which is a key component of many fertilizers.

Coal is an important natural resource in India that is used for a wide range of industrial and domestic purposes. The country's significant coal reserves make it an important player in the global energy market, and the Indian government has taken steps to increase coal production and improve the efficiency of coal-fired power plants in recent years.

Lithological interpretation of sedimentary rocks can be aided by visual inspection and hand specimen examination, which involves looking at the size, shape, and texture of the grains or particles. Petrographic microscope can be used to examine thin sections of the rocks and to identify the minerals present. Various analytical techniques such as X-ray diffraction, scanning electron microscopy, and stable isotope analysis can also be used to further interpret the lithology and depositional environment of the rocks. Understanding the lithology of sedimentary rocks is important in geology, as it provides insights into the Earth's geologic history, past climate, and the depositional environment in which the rocks were formed.

Lithological Interpretation of Metamorphic Rocks

Lithological interpretation of metamorphic rocks involves the identification and description of the different types of rocks that have undergone changes in texture, mineralogy, or chemical composition due to the effects of heat, pressure, and/or chemical activity. Metamorphic rocks can be classified based on their texture, mineralogy, and degree of metamorphism. Here are some of the common types of metamorphic rocks and their characteristics:

1. Slate

Slate is a fine-grained metamorphic rock that is characterized by a distinct foliation. It is typically formed from the metamorphism of shale or mudstone.

slate is a type of metamorphic rock that is formed from the alteration of shale or mudstone under low-grade regional metamorphism. The process of metamorphism involves the alteration of pre-existing rocks under heat and pressure conditions, resulting in the development of new minerals and textures.

Slate is characterized by its foliated texture, which means it has distinct layers or planes of minerals that are aligned parallel to each other. The layers of slate are typically very fine-grained and can be split easily into thin sheets, making slate a valuable material for roofing tiles and other construction uses.

Slate is primarily composed of the mineral mica, which gives it its distinctive sheen and smooth surface. However, other minerals such as quartz, feldspar, and chlorite can also be present in slate.

In geology, slate is important because it provides information about the conditions of regional metamorphism, such as temperature and pressure, that existed during its formation. The presence of slate in an area can also indicate the potential for other types of metamorphic rocks to be present, as well as the potential for mineral deposits to be found.

Slate is an important metamorphic rock in both construction and geology due to its unique properties and valuable insights into the processes that shape the Earth's crust.

Slate in India: Slate is a fine-grained, metamorphic rock that is formed from the compression of clay and shale over millions of years. In India, slate is found in many parts of the country, including in the states of Himachal Pradesh, Uttarakhand, Andhra Pradesh, and Maharashtra.

Uses: Here are some common uses of slate in India:

- Roofing: Slate is used as a roofing material in many parts of India. Its durability, resistance to weathering, and aesthetic appeal make it a popular choice for roofing homes, commercial buildings, and other structures.
- Flooring: Slate is used as a flooring material in homes, offices, and other buildings. Its smooth texture and natural color make it a popular choice for interior and exterior flooring.
- Cladding: Slate is used as a cladding material for walls and other surfaces. Its natural beauty and durability make it a popular choice for cladding the exterior of buildings and other structures.
- Landscaping: Slate is used for landscaping purposes in gardens, parks, and other outdoor spaces. It is used to create paths, retaining walls, and other decorative features.
- Crafts: Slate is sometimes used for crafts and art projects. Its fine texture and ability to hold detailed carvings make it a popular choice for sculptures, engravings, and other artistic works.

Slate is a versatile and durable natural resource in India that is used for a wide range of industrial, commercial, and artistic purposes. Its unique texture and natural beauty make it a popular choice for roofing, flooring, cladding, landscaping, and crafts.

2. Schist

Schist is a medium- to coarse-grained metamorphic rock that is characterized by a well-developed foliation and a high degree of mineral segregation. It is typically formed from the metamorphism of shale, mudstone, or basalt.

Schist is a type of metamorphic rock that is characterized by its foliated texture, which means it has distinct layers or bands of minerals. Schist forms from other rocks, such as shale or mudstone, that are subjected to high heat and pressure deep within the Earth's crust. The minerals in the original rock recrystallize and align themselves in bands or layers, giving schist its distinctive texture.

Schist can come in a range of colors and is often used as a decorative stone in construction and landscaping. Some common types of schist include mica schist, garnet schist, and hornblende schist. Mica schist, for example, is a type of schist that contains large amounts of mica, which gives it a shimmering appearance. Schist can also contain other minerals such as quartz, feldspar, and amphibole.

Schist is an important rock in geology because it provides information about the tectonic processes that shape the Earth's crust. The presence of schist can indicate areas of high pressure and temperature, such as subduction zones or collision zones between tectonic plates. It can also be a valuable source of minerals such as garnet, mica, and graphite.

Schist in India: Schist is a type of metamorphic rock that is formed from the alteration of other rocks through heat and pressure over time. In India, schist is found in many parts of the country, including the states of Uttarakhand, Himachal Pradesh, Jammu and Kashmir, and Maharashtra.

Uses: Schist is a type of metamorphic rock that has various uses due to its unique properties. Here are some common uses of schist:

- Building materials: Schist is used as a building material in construction. It is used for flooring, roofing, cladding, and walls due to its durability, strength, and resistance to weathering. Schist can be cut into thin sheets or used in thicker slabs for structural purposes.
- Landscaping: Schist is used for landscaping projects like garden paths, retaining walls, and decorative features. Its texture and color give a natural look to the outdoor spaces.
- Decorative uses: Schist is also used in interior design for decorative purposes like flooring, wall cladding, and decorative tiles. It gives a unique and natural look to the interiors.
- Industrial uses: Schist can be used in the production of abrasives and other industrial materials due to its hardness and durability. It is also used for manufacturing tiles, slabs, and countertops.
- Artistic uses: Schist is used for artistic purposes like sculptures, carvings, and engravings due to its unique texture and color.
- Metallurgical uses: Schist is used as a source of raw materials for iron and steel production.

Schist is a versatile material with various uses in construction, landscaping, industry, and art. Its unique texture, durability, and resistance to weathering make it a popular choice for many different applications.

3. Gneiss

Gneiss is a coarse-grained metamorphic rock that is characterized by a banded texture and a high degree of mineral segregation. It is typically formed from the metamorphism of granite or other igneous rocks.

Gneiss is a type of metamorphic rock that is formed by the alteration of pre-existing rocks under high pressure and temperature conditions. It is characterized by its banded or layered appearance, which is a result of the segregation of different minerals into distinct bands or layers during the metamorphic process.

Gneiss is typically composed of minerals such as quartz, feldspar, and mica, and may also contain other minerals such as garnet, amphibole, and pyroxene. The color and composition of gneiss can vary depending on the type of pre-existing rock that it was formed from and the specific metamorphic conditions it underwent.

Gneiss is often used in construction as a decorative building stone because of its unique texture and appearance. It is also commonly used as a source of crushed stone for road construction and as a material for landscaping.

In geology, gneiss is important because it provides information about the geological processes that have shaped the Earth's crust. Its formation is often associated with regional metamorphism, which can occur during tectonic processes such as mountain building and continental collision. Gneiss is also important for understanding the history and evolution of the Earth's continents, as it is commonly found in ancient continental shields and other areas of exposed basement rock.

Gneiss in India: Gneiss is a metamorphic rock that is formed through the process of regional metamorphism, which occurs under high pressure and temperature conditions over a large area. In India, gneiss formation can be found in many regions, particularly in the Precambrian shield areas of the country.

The formation of gneiss in India is closely related to the geological history of the region. During the Precambrian era, India was part of a larger landmass called Gondwana, which was later split into smaller continents. The Indian subcontinent collided with the Eurasian plate around 50 million years ago, causing significant tectonic activity in the region. This activity, along with the effects of weathering and erosion, led to the formation of gneiss.

In India, gneiss can be found in many states, including Kerala, Karnataka, Tamil Nadu, Andhra Pradesh, Rajasthan, and Madhya Pradesh. These rocks are often used as building materials, road construction materials, and decorative stones due to their durability and attractive appearance.

Gneiss is a valuable rock type due to its properties, such as resistance to weathering and erosion, attractive appearance, and durability. The formation of gneiss in India is a significant geological process that has contributed to the development of the country's geology and natural resources.

Uses: Gneiss is a type of metamorphic rock that has several uses due to its unique properties. Here are some common uses of gneiss:

Building materials: Gneiss is used as a building material in construction due to its durability, strength, and resistance to weathering. It is used for flooring, roofing, cladding, and walls.

Landscaping: Gneiss is used for landscaping projects like garden paths, retaining walls, and decorative features. Its texture and color give a natural look to the outdoor spaces.

Decorative uses: Gneiss is also used in interior design for decorative purposes like flooring, wall cladding, and decorative tiles. It gives a unique and natural look to the interiors.

Industrial uses: Gneiss can be used in the production of abrasives and other industrial materials due to its hardness and durability. It is also used for manufacturing tiles, slabs, and countertops.

Artistic uses: Gneiss is used for artistic purposes like sculptures, carvings, and engravings due to its unique texture and color.

Metallurgical uses: Gneiss is used as a source of raw materials for iron and steel production.

Gneiss is a versatile material with various uses in construction, landscaping, industry, and art. Its unique texture, durability, and resistance to weathering make it a popular choice for many different applications.

4. Marble

Marble is a metamorphic rock that is composed of calcium carbonate. It is typically formed from the metamorphism of limestone or dolomite.

Marble is a type of metamorphic rock that is formed from the alteration of limestone or dolomite rock under high pressure and temperature conditions. Marble is known for its distinctive appearance, which is characterized by its smooth, polished surface and unique veining patterns.

Marble is composed primarily of calcium carbonate, which is the same mineral found in limestone and many other types of sedimentary rocks. However, the high pressure and temperature conditions during metamorphism cause the calcium carbonate to recrystallize and form interlocking crystals of calcite or dolomite, giving marble its characteristic texture and appearance.

Marble is prized for its beauty and is commonly used as a decorative stone in construction and sculpture. It has been used for centuries in buildings, monuments, and works of art, including famous examples such as the Taj Mahal in India and the Parthenon in Greece.

In addition to its aesthetic value, marble is also valued for its durability and resistance to weathering and erosion. It is often used as a building material in areas with a high risk of seismic activity or other natural disasters because of its ability to withstand damage.

Marble is an important rock in geology and the construction industry due to its unique properties and aesthetic appeal. Its formation provides valuable insights into the processes that shape the Earth's crust, and its use in architecture and sculpture has left an enduring mark on human history and culture.

Marble in India: Marble is a type of metamorphic rock that is composed of recrystallized carbonate minerals, typically calcite or dolomite. In India, marble is widely found in the states of Rajasthan, Gujarat, and Andhra Pradesh. Some of the famous varieties of marble found in India are Makrana marble, Ambaji marble, Abu black marble, and Udaipur green marble.

Uses: Marble has been used in India for centuries for its beauty, durability, and versatility. It is a popular material for building construction, sculptures, decorative items, and flooring. Here are some common uses of marble in India:

Construction: Marble is used in construction as a decorative stone for cladding, flooring, and walls. It is also used for creating columns, pillars, and arches due to its strength and durability.

Sculptures: Marble is a popular material for creating sculptures due to its ability to be carved and shaped easily. Many famous sculptures in India, such as the Taj Mahal, are made of marble.

Decorative items: Marble is used to make a variety of decorative items like vases, lamps, bowls, and statues. These items are popular for home decor and gifting purposes.

Tombstones: Marble is a popular material for making tombstones and monuments due to its durability and resistance to weathering.

Tabletops: Marble is used to make tabletops, kitchen countertops, and bathroom vanity tops. It is popular for these applications due to its resistance to heat and stains.

Marble is a popular material in India due to its beauty, durability, and versatility. Its uses range from construction to decorative items, sculptures, and more.

5. Quartzite

Quartzite is a metamorphic rock that is composed of quartz. It is typically formed from the metamorphism of sandstone.

Quartzite is a type of metamorphic rock that is formed from the recrystallization of sandstone under high temperature and pressure. Quartzite is composed primarily of quartz, a mineral that is found in many types of rocks, including granite and sandstone.

Quartzite is known for its hardness, durability, and resistance to weathering and erosion. It is often used as a building material and decorative stone because of its unique texture and color variations. Quartzite can come in a range of colors, including white, gray, pink, red, and green, and its appearance can be characterized by its sugary or granular texture.

Quartzite is also valued for its resistance to heat, which makes it a popular choice for use in kitchens and bathrooms. It is also commonly used as a material for landscaping, including for walkways, walls, and outdoor living spaces.

In geology, quartzite is important because it provides insights into the processes that shape the Earth's crust. The formation of quartzite involves the recrystallization of sandstone, which can occur during regional metamorphism associated with tectonic activity such as mountain building. The presence of quartzite in an area can provide information about the geological history and tectonic processes that have occurred there.

Quartzite is an important rock in both geology and the construction industry due to its unique properties and versatility. Its hardness and durability make it a valuable building material, while its formation provides valuable insights into the processes that shape the Earth's crust.

Quartzite in India: Quartzite is a type of metamorphic rock that is composed of quartz grains that have been recrystallized and fused together. In India, quartzite is mainly found in the states of Rajasthan, Madhya Pradesh, and Chhattisgarh. Some of the famous varieties of quartzite found in India are Bhilwara Green quartzite, Deoli Green quartzite, Himachal White quartzite, and Pink Quartzite.

Uses: Quartzite has a wide range of uses due to its unique properties. Here are some common uses of quartzite in India:

Construction: Quartzite is used as a decorative stone for cladding, flooring, and walls. It is also used for creating steps, pavements, and landscaping due to its durability and resistance to weathering.

Roofing: Quartzite is used as a roofing material due to its ability to resist weathering, erosion, and fading. It is also fire-resistant and provides insulation against heat and cold.

Industrial uses: Quartzite is used in the production of silica bricks, refractory materials, and glass. It is also used as an abrasive in sandblasting, grinding, and polishing due to its hardness.

Decorative items: Quartzite is used to make a variety of decorative items like vases, lamps, bowls, and statues. These items are popular for home decor and gifting purposes.

Landscaping: Quartzite is used for creating garden paths, retaining walls, and decorative features in landscaping. Its texture and color give a natural look to the outdoor spaces.

Quartzite is a versatile material with various uses in construction, roofing, industry, and art. Its unique properties like durability, resistance to weathering, and hardness make it a popular choice for many different applications in India.

Lithological interpretation of metamorphic rocks can be aided by visual inspection and hand specimen examination, which involves looking at the texture and mineralogy of the rocks. Petrographic microscope can be used to examine thin sections of the rocks and to identify the minerals present. Various analytical techniques such as X-ray diffraction and stable isotope analysis can also be used to further interpret the lithology and the conditions under which the rocks were metamorphosed. Understanding the lithology of metamorphic rocks is important in geology, as it provides insights into the Earth's geologic history, tectonic activity, and the conditions under which the rocks were metamorphosed.

Important Questions

1. What are the key characteristics used to differentiate igneous rocks from sedimentary and metamorphic rocks?
2. How does the mineral composition of igneous rocks differ from that of sedimentary rocks?
3. What are the primary processes responsible for the formation of sedimentary rocks?
4. How do the textures of sedimentary rocks vary depending on the depositional environment?
5. What are the main types of metamorphic rocks and how do they form?
6. What are the characteristic features that allow the identification of metamorphic rocks?

7. How do contact and regional metamorphism differ in terms of their geological settings and rock textures?
8. How do sedimentary rocks provide clues about past environmental conditions?
9. What are the different classifications of igneous rocks based on their mineral composition and texture?
10. How does the geological history of a region influence the information and distribution of sedimentary, igneous, and metamorphic rocks?

3

Geological Classification of India Major Rock Type

India is a geologically diverse country with a wide variety of rock types. The major rock types found in India can be broadly classified into the following categories:

Igneous Rocks: These rocks are formed from the solidification of molten magma. They can be further divided into two types:

a. Plutonic (Intrusive) Igneous Rocks: Examples include granite, diorite, and gabbro. These rocks are formed deep within the Earth's crust and have a coarse-grained texture.

b. Volcanic (Extrusive) Igneous Rocks: Examples include basalt, andesite, and rhyolite. These rocks are formed on the Earth's surface from volcanic eruptions and have a fine-grained texture.

Sedimentary Rocks: These rocks are formed by the deposition and consolidation of sediments over time. They can be further divided into various types based on their composition and mode of formation:

a. Clastic Sedimentary Rocks: Examples include sandstone, shale, and conglomerate. These rocks are formed from the accumulation and lithification of fragments of pre-existing rocks.

b. Chemical Sedimentary Rocks: Examples include limestone, dolomite, and rock salt. These rocks are formed from the precipitation of minerals from water bodies.

c. Organic Sedimentary Rocks: Examples include coal and petroleum. These rocks are formed from the accumulation and decomposition of organic matter.

Metamorphic Rocks: These rocks are formed from the transformation of pre-existing rocks due to high temperatures, pressures, and/or chemical reactions. They can be further divided into various types:

a. Regional Metamorphic Rocks: Examples include gneiss, schist, and slate. These rocks are formed over large areas as a result of regional tectonic processes.

b. Contact Metamorphic Rocks: Examples include marble and quartzite. These rocks are formed near igneous intrusions or along faults due to localized heat and pressure.

It's important to note that the distribution and occurrence of these rock types can vary across different regions of India, and there may be additional rock types found in specific geological formations or regions.

Igneous Rocks in India

India has a diverse range of igneous rocks distributed across its geological formations. Here are some prominent igneous rock types found in different regions of India:

Granite: Granite is a common plutonic igneous rock found in several regions of India. It is composed mainly of quartz, feldspar, and mica minerals. Some notable granite regions in India include the Aravalli Range in Rajasthan, the Eastern Ghats in Odisha and Andhra Pradesh, and parts of Karnataka.

Basalt: Basalt is a prevalent volcanic extrusive rock found in many parts of India. It is dark in color and composed mainly of plagioclase feldspar and pyroxene minerals. Basalt formations are widespread, including the Deccan Plateau in central India, the Western Ghats, parts of Maharashtra, Gujarat, and the Andaman and Nicobar Islands.

Dolerite: Dolerite is a medium-grained intrusive igneous rock similar to basalt. It is commonly found as dykes or sills cutting across other rocks. Dolerite occurrences can be seen in various regions of India, including the Deccan Plateau, the Eastern Ghats, and parts of Maharashtra, Karnataka, and Tamil Nadu.

Gabbro: Gabbro is a coarse-grained intrusive igneous rock composed mainly of calcium-rich plagioclase feldspar and pyroxene minerals. Gabbroic rocks can be found in the Aravalli Range in Rajasthan, parts of Gujarat, and some regions of Tamil Nadu.

Rhyolite: Rhyolite is a fine-grained volcanic extrusive rock rich in silica content. It typically has a light-colored appearance due to the abundance of quartz and feldspar minerals. Rhyolite occurrences can be found in areas such as the Western Ghats, parts of Maharashtra, Karnataka, and the Andaman and Nicobar Islands.

Andesite: Andesite is an intermediate volcanic rock with a composition between basalt and rhyolite. It commonly occurs in volcanic arcs and subduction zones. Andesite formations can be found in the Eastern Ghats, parts of Arunachal Pradesh, and the Andaman and Nicobar Islands.

These are just a few examples of the igneous rocks found in India. The country's geological diversity ensures the presence of numerous other igneous rock types, each with its unique characteristics and distribution patterns.

Sedimentary Rocks in India

India is rich in various sedimentary rock formations that have been formed over millions of years. Here are some prominent sedimentary rock types found in different regions of India:

Gondwana Sedimentary Rocks: The Gondwana Group of sedimentary rocks is one of the most significant formations in India. These rocks were deposited during the Permian and Triassic periods and are widely distributed across central and eastern India. The Gondwana rocks consist of various types, including sandstone, shale, and coal.

Vindhyan Sedimentary Rocks: The Vindhyan Supergroup is another major sedimentary rock formation in India, extending across central India. These rocks were deposited during the Proterozoic era and comprise sandstone, shale, limestone, and dolomite. The Vindhyan rocks are renowned for their fossil-rich nature.

Cuddapah Sedimentary Rocks: The Cuddapah Basin in southern India is known for its Cuddapah Supergroup of sedimentary rocks. These rocks were formed during the Proterozoic era and consist of shale, sandstone, and limestone. They are particularly well-exposed in the regions of Andhra Pradesh and Telangana.

Siwalik Sedimentary Rocks: The Siwalik Group is a significant sedimentary rock formation found along the foothills of the Himalayas in northern India. These rocks are of Miocene and Pliocene age and comprise sandstone, shale, conglomerate, and clay. The Siwalik rocks are important for their fossil record, including mammal fossils.

Indo-Gangetic Alluvium: The Indo-Gangetic plain in northern India is characterized by extensive alluvial deposits. These sediments have been brought by major rivers such as the Ganges, Brahmaputra, and Indus over thousands of years. The alluvial deposits consist of sand, silt, clay, and gravel and are highly fertile, supporting agricultural activities in the region.

Karst Limestone: Karst limestone formations can be found in various parts of India, notably in regions such as Meghalaya, parts of Gujarat, and parts of Madhya Pradesh. These limestone formations are characterized by their unique topography, featuring caves, sinkholes, and underground rivers.

These are some of the prominent sedimentary rock types and formations found in India. The country's geological history and diverse landscape have resulted in the deposition and preservation of a wide range of sedimentary rocks across different regions.

Metamorphic Rocks in India

India has a diverse range of metamorphic rocks that have been formed through the transformation of pre-existing rocks under high temperature, pressure, and/or chemical activity. Here are some notable metamorphic rock types found in different regions of India:

Gneiss: Gneiss is a foliated metamorphic rock characterized by alternating bands of light and dark minerals. It is widespread in India and can be found in regions such as the Aravalli Range in Rajasthan, the Eastern Ghats, parts of Karnataka and Tamil Nadu, and the Himalayan region.

Schist: Schist is another foliated metamorphic rock commonly found in India. It has a medium to coarse-grained texture and exhibits a schistosity or layered structure. Schist formations can be seen in various regions, including the Aravalli Range, the Eastern Ghats, parts of Maharashtra and Odisha, and the Himalayas.

Slate: Slate is a fine-grained metamorphic rock that forms from the low-grade metamorphism of shale or mudstone. It is characterized by its ability to split into thin, flat sheets. Slate deposits can be found in regions such as Himachal Pradesh, Uttarakhand, and parts of the Northeastern states.

Marble: Marble is a metamorphic rock formed from the recrystallization of limestone or dolomite. It is known for its distinctive veined or banded appearance and is widely used as a decorative stone. Marble occurrences can be found in several regions, including Rajasthan (Makrana and Udaipur), Gujarat, Madhya Pradesh, and parts of Maharashtra.

Quartzite: Quartzite is a metamorphic rock formed from the recrystallization of quartz-rich sandstone. It is extremely hard and durable and often used as a construction material. Quartzite deposits can be found in regions such as the Aravalli Range, the Vindhyan Basin, parts of Maharashtra, and parts of Tamil Nadu.

Amphibolite: Amphibolite is a metamorphic rock composed mainly of amphibole minerals, typically hornblende. It has a coarse-grained texture and can often be found associated with other metamorphic rocks. Amphibolite occurrences can be seen in the Aravalli Range, the Eastern Ghats, parts of Odisha, and the Himalayas.

These are just a few examples of the metamorphic rock types found in India. The country's varied geological history and the presence of tectonic activity have led to the formation of numerous metamorphic rocks across different regions.

Important Questions

1. What are the major rock types found in India?
2. How do igneous rocks contribute to India's geological classification?
3. What are the main characteristics of sedimentary rocks in India?
4. How do metamorphic rocks play a role in India's geological composition?
5. Which region in India is known for its significant granite deposits?
6. What are the major rock formations found in Western Ghats of India?
7. How has the Deccan Traps volcanic province influenced India's geological classification?
8. What are the key features of the Aravalli Range in terms of rock types?
9. How do the Himalayas contribute to India's geological diversity?
10. What is the significance of the Dharwar Supergroup in India's geological history?

4

Interpretation of Drainage Patterns

Interpretation of Drainage Patterns Through Aerial Photographs

Aerial photographs can provide a useful tool for interpreting the drainage patterns of a landscape. Drainage patterns are formed by the natural movement of water across the land and the way in which it flows from high to low areas, creating channels and streams that eventually converge into larger rivers and bodies of water. Different types of drainage patterns can be identified through aerial photographs, which can help to understand the underlying geology and topography of the landscape.

Here are some of the different drainage patterns that can be identified through aerial photographs:

1. Dendritic pattern

This pattern resembles the branches of a tree and is the most common type of drainage pattern. It occurs in areas where the underlying rock or soil is uniform and has no preferred direction of erosion. The streams converge into larger rivers, which may flow in a particular direction.

Dendritic patterns are common in river systems, where the tributaries of a main river resemble the branches of a tree. In India, there are many river systems that exhibit dendritic patterns, including:

The Ganges River: The Ganges is one of the largest and most important rivers in India, and it exhibits a dendritic pattern in its upper reaches in the Himalayan region.

The Brahmaputra River: The Brahmaputra is another major river in India that flows through the Himalayan region, and it exhibits a dendritic pattern in its upper reaches.

The Godavari River: The Godavari is a major river in southern India, and it exhibits a dendritic pattern in its upper reaches in the Western Ghats.

The Krishna River: The Krishna is another major river in southern India, and it exhibits a dendritic pattern in its upper reaches in the Western Ghats.

The Mahanadi River: The Mahanadi is a major river in eastern India, and it exhibits a dendritic pattern in its upper reaches in the Chhattisgarh region.

These dendritic patterns are shaped by the underlying geology and topography of the region, as well as the erosion and sedimentation processes that occur over time. Understanding the dendritic patterns of rivers can be important for managing water resources, predicting flooding and erosion, and understanding the natural history of the region.

2. Radial pattern

This pattern forms when streams flow from a central peak or high point, such as a volcano. The streams flow in a radial pattern outward from the peak, creating a spoke-like appearance.

There is no such thing as a radial pattern of rivers in India. The radial pattern typically refers to a design concept used in urban planning, where streets and roads are arranged in a circular or radial pattern around a central point or landmark. However, when it comes to rivers in India, there is no known pattern that is strictly radial in nature.

Rivers in India flow in various directions and follow the contours of the landscape. For example, the Ganges River flows roughly from the northwest to the southeast, while the Brahmaputra River flows from the northeast to the southwest. The Godavari River, on the other hand, flows from west to east before draining into the Bay of Bengal.

However, it is worth noting that rivers in India often exhibit complex patterns, with many tributaries branching out in different directions and joining larger rivers at various points along their course. This creates a dense network of rivers that crisscrosses the country, providing vital water resources for agriculture, industry, and domestic use.

While there is no radial pattern of rivers in India, the complex network of rivers and tributaries plays a vital role in the country's ecology, economy, and culture.

3. Rectangular pattern

This pattern is characterized by a series of interconnected streams that flow at right angles to each other. It is commonly found in areas with a strong joint pattern in the underlying rock, which causes the streams to follow the straight lines created by the joints.

The rectangular pattern of rivers in India refers to the network of rivers that flow in a roughly north-south or east-west direction across the Indian subcontinent, forming a rectangular or grid-like pattern on the map. This pattern is the result of the geological history of the region and the tectonic forces that have shaped the landscape over millions of years.

The two major river systems in India, the Ganges-Brahmaputra-Meghna and the Indus, flow in a roughly north-south direction, while many of their tributaries flow in an east-west direction, creating a rectangular or grid-like pattern on the map. This pattern is also influenced by the regional climate, with monsoonal rains causing the rivers to shift their courses and create new channels over time.

The rectangular pattern of rivers in India has played an important role in the development of the country's civilization and economy. The rivers have provided water for agriculture, transportation for goods and people, and a source of energy through hydroelectric power generation. However, the rivers have also been a source of conflict, as different regions and states have competed for control of water resources and management of river systems.

The rectangular pattern of rivers in India is a distinctive feature of the country's geography and has had a significant impact on its history, culture, and economy.

4. Trellis pattern

This pattern forms when streams flow parallel to each other in a linear pattern, with smaller streams flowing into larger ones at right angles. It is commonly found in areas with alternating layers of hard and soft rock, which erode at different rates.

The trellis pattern of rivers in India refers to a distinct pattern of river drainage that is characterized by a main river channel that is joined by many smaller tributaries at right angles. This pattern is often found in regions with alternating layers of hard and soft rock, where the softer rocks are eroded more easily by water, creating valleys that are perpendicular to the main river channel.

One of the most prominent examples of the trellis pattern of rivers in India is found in the Deccan Plateau region, particularly in the state of Maharashtra. The Godavari River, which flows through Maharashtra and several other states, is one of the largest rivers in India with a trellis pattern. The tributaries of the Godavari River, such as the Manjira and the Penganga, flow in a roughly east-west direction and join the main river channel at right angles.

The trellis pattern of rivers in India has important implications for water resource management and development. The main river channels and tributaries in these systems often provide different levels of water availability, quality, and accessibility, which can affect agriculture, industry, and human settlements in the surrounding areas.

Trellis pattern of rivers in India is a unique and important feature of the country's geography that has shaped its ecology, economy, and culture over thousands of years.

5. Parallel pattern

This pattern occurs when streams flow parallel to each other, usually due to a steep gradient in the landscape or the presence of a resistant layer of rock. The streams may converge into larger rivers at the bottom of the slope.

The parallel pattern of rivers in India refers to a distinct pattern of river drainage in which multiple rivers flow parallel to each other, usually separated by ridges or higher elevations. This pattern is often found in regions with parallel ridges and valleys, where the rivers have eroded the soft rock between the ridges, creating a network of parallel river channels.

One of the most prominent examples of the parallel pattern of rivers in India is found in the region of peninsular India, particularly in the Western Ghats mountain range. The rivers that flow from the Western Ghats, such as the Krishna, Cauvery, and Godavari, flow in parallel to each other, separated by the mountain ridges and plateaus.

The parallel pattern of rivers in India has important implications for water resource management and development. The rivers in these systems often provide different levels of water availability, quality, and accessibility, which can affect agriculture, industry, and human settlements in the surrounding areas. Moreover, the parallel river channels create unique ecological habitats, supporting a variety of flora and fauna.

The parallel pattern of rivers in India is a unique and important feature of the country's geography that has shaped its ecology, economy, and culture over thousands of years.

By examining the aerial photographs of a landscape, one can identify the dominant drainage pattern and use it to gain insights into the underlying geology and topography. This information can be useful for a range of applications, such as land use planning, environmental management, and engineering design.

Satellite Images Interpretation of Glacial Landforms

Interpreting glacial landforms from satellite images can be challenging, but with careful examination and understanding of glacial processes, it is possible to identify many glacial features. Some of the common glacial landforms that can be observed in satellite images include:

1. Moraines

These are ridges of rock and sediment that mark the edges or former extent of a glacier. They can be lateral (running parallel to the valley sides), medial (running down the center of a valley), or terminal (at the end of a glacier).

Moraines are glacial landforms that are created by the deposition of rocks, soil, and debris carried by glaciers. In India, moraines are found in the high-altitude mountain regions that have been glaciated in the past.

The most prominent moraines in India are found in the Himalayas, where the world's highest mountain peaks are located. Glaciers that originated from the Himalayas have left behind a wide variety of moraines, including lateral moraines, medial moraines, terminal moraines, and recessional moraines.

Lateral moraines are formed along the sides of glaciers and are created when debris is eroded from the surrounding mountains and deposited onto the glacier. Medial moraines are formed when two glaciers merge, and the debris carried by each glacier creates a new moraine in the middle of the merged glacier. Terminal moraines are formed at the end of glaciers and mark the farthest point reached by the glacier. Recessional moraines are formed as glaciers retreat, leaving behind piles of debris that were previously carried by the glacier.

The moraines in India are not only significant in terms of their geological formation but also for their ecological and cultural importance. These glacial landforms are an important source of freshwater for local communities and support a unique ecosystem of plants and animals that have adapted to the harsh mountain environment. Moreover, many of the moraines in India are associated with religious and cultural sites that have great significance to the people who live in the surrounding regions.

Moraines are a significant glacial landform in India that have played an important role in shaping the country's landscape, ecology, and culture.

2. Drumlins

These are elongated hills that are typically streamlined in the direction of ice flow. They are often found in groups and are thought to have been formed by the deposition of sediment beneath a moving glacier.

Drumlins are glacial landforms that are characterized by a smooth, elongated, and steep-sided hill or ridge, typically composed of glacial till or other unconsolidated sediments. These landforms are formed by the action of glaciers and are usually found in regions that have experienced past glaciation.

In India, there is limited evidence of drumlins being present due to the limited glaciation that has occurred in the country. Most of the glacial landforms in India are found in the Himalayas, and these are primarily comprised of moraines, cirques, and U-shaped valleys.

However, recent studies have indicated the presence of some glacial landforms that are similar to drumlins in the Ladakh region of Jammu and Kashmir. These features are found in the form of elongated, streamlined hills that are aligned with the direction of ice flow. These hills are composed of glacial till and are similar in shape to drumlins, though they are not as pronounced as the drumlins found in other parts of the world.

Despite being relatively rare in India, the presence of drumlin-like landforms in the Ladakh region provides valuable insights into the glacial history of the Indian subcontinent. It highlights the need for further research into the extent and nature of past glaciation in the region and the potential impacts of future climate change on these glacial landforms.

While the presence of drumlins in India is limited, they remain an important glacial landform that contributes to our understanding of the country's geology and past climate history.

3. Eskers

These are long, winding ridges of sediment that were deposited by meltwater streams flowing beneath a glacier. They often form sinuous ridges that can be easily recognized in satellite imagery.

Eskers are long, winding ridges of sediment that are typically formed in the subglacial environment by the deposition of meltwater and sediment from retreating glaciers. While India is not typically associated with glacial landforms, there is evidence of eskers in the Ladakh region of Jammu and Kashmir, which is located in the Himalayan mountain range.

The eskers in Ladakh are believed to have been formed during the Pleistocene epoch, which lasted from about 2.6 million to 11,700 years ago, when large parts of the Indian subcontinent were covered by ice sheets and glaciers. These eskers are believed to have been formed by meltwater streams flowing beneath the glaciers, depositing sediment as the glaciers retreated.

The Ladakh eskers are important glacial landforms because they provide valuable insights into the glacial history of the Indian subcontinent. They offer clues about the extent and nature of past glaciation in the region and can help scientists understand how the Himalayan mountain range was formed. In addition, eskers are important sources of groundwater and are often used as sites for water extraction in areas with limited water resources.

While eskers are not a common glacial landform in India, their presence in the Ladakh region provides important information about the country's geological history and can help us understand the impact of past and future climate change on the region.

4. Kames

These are small hills or mounds of sediment that were deposited by glacial meltwater. They are often found in groups and can be easily identified in satellite imagery.

Kames are small, cone-shaped or oval mounds of sand and gravel that are often found near the end of a glacier or in regions where glacial meltwater is abundant. While India is not typically associated with glacial landforms, there is some evidence of kames in the Ladakh region of Jammu and Kashmir, which is located in the Himalayan mountain range.

The kames in Ladakh are believed to have been formed during the Pleistocene epoch, which lasted from about 2.6 million to 11,700 years ago, when large parts of the Indian subcontinent were covered by ice sheets and glaciers. They are thought to have been formed by meltwater streams flowing out of the glaciers and depositing sediment in cone-shaped or oval mounds.

The Ladakh kames are important glacial landforms because they provide valuable insights into the glacial history of the Indian subcontinent. They offer clues about the extent and nature of past glaciation in the region and can help scientists understand how the Himalayan mountain range was formed. In addition, kames can be important sources of groundwater and can serve as sites for water extraction in areas with limited water resources.

While kames are not a common glacial landform in India, their presence in the Ladakh region provides important information about the country's geological history and can help us understand the impact of past and future climate change on the region.

5. Outwash plains

These are broad, flat areas of sediment that were deposited by meltwater streams as they flowed away from a glacier. They can be recognized in satellite imagery as large, flat areas with relatively low relief.

Outwash plains are flat, gently sloping plains of sediment that are formed by meltwater streams flowing out of a glacier. These landforms are typically composed of sand, gravel, and other glacial sediments that have been sorted and deposited by the meltwater streams. While India is not typically associated with glacial landforms, there is some evidence of outwash plains in the Ladakh region of Jammu and Kashmir, which is located in the Himalayan mountain range.

The outwash plains in Ladakh are believed to have been formed during the Pleistocene epoch, which lasted from about 2.6 million to 11,700 years ago, when large parts of the Indian subcontinent were covered by ice sheets and glaciers. They are thought to have been formed by meltwater streams flowing out of the glaciers and depositing sediment in flat, gently sloping plains.

The Ladakh outwash plains are important glacial landforms because they provide valuable insights into the glacial history of the Indian subcontinent. They offer clues about the extent and nature of past glaciation in the region and can help scientists understand how the Himalayan mountain range was formed. In addition, outwash plains can be important sources of groundwater and can serve as sites for water extraction in areas with limited water resources.

While outwash plains are not a common glacial landform in India, their presence in the Ladakh region provides important information about the country's geological history and can help us understand the impact of past and future climate change on the region.

6. Cirques

These are bowl-shaped depressions that form at the head of a glacier. They can be recognized in satellite imagery as amphitheater-shaped valleys with steep walls.

There are several cirques or glacial landforms in India, particularly in the northern region of the country, where the Himalayan mountain range is located. Some of the notable cirques in India are:

Gangotri Glacier Cirque: This is one of the largest and most famous cirques in India. It is located in the Uttarkashi district of Uttarakhand and is the source of the Ganges River.

Siachen Glacier Cirque: This cirque is located in the Karakoram Range of the Himalayas in the Ladakh region of Jammu and Kashmir. It is the world's second-longest glacier outside of the polar regions.

Satopanth Glacier Cirque: This cirque is situated in the Chamoli district of Uttarakhand and is known for its stunning views of the surrounding peaks.

Milam Glacier Cirque: This cirque is located in the Pithoragarh district of Uttarakhand and is one of the largest glaciers in the region.

Gangri Glacier Cirque: This cirque is situated in the Kinnaur district of Himachal Pradesh and is known for its unique geology and stunning landscape.

These cirques are not only important from a geological and scientific perspective but are also popular tourist destinations for adventure enthusiasts and nature lovers.

7. Hanging valleys

These are small valleys that are left above the level of a larger valley as a glacier retreats. They can be recognized in satellite imagery as smaller valleys that end abruptly at a steep slope.

Hanging valleys are another type of glacial landform that can be found in India. These valleys are formed when smaller glaciers join a larger glacier and erode the surrounding landscape at a different rate, creating a valley that is "hanging" above the main valley. Some of the hanging valleys in India include:

Panpatia Hanging Valley: This hanging valley is located in the Chamoli district of Uttarakhand and is known for its steep slopes and scenic views of the surrounding peaks.

Suralaya Hanging Valley: This hanging valley is located in the Lahaul and Spiti district of Himachal Pradesh and is known for its unique geological formations and stunning views of the Himalayan mountains.

Sangla Hanging Valley: This hanging valley is situated in the Kinnaur district of Himachal Pradesh and is known for its lush vegetation and picturesque waterfalls.

Kangla Glacier Hanging Valley: This hanging valley is located in the Ladakh region of Jammu and Kashmir and is known for its rugged terrain and stunning glacial landscape.

These hanging valleys are not only important from a geological and scientific perspective, but also serve as popular tourist destinations for trekking and adventure activities.

In addition to these landforms, other glacial features such as crevasses, seracs, and icefalls can also be observed in satellite imagery. Understanding these features can help researchers better understand past and present glacial processes and how they may be impacted by climate change.

Satellite Images Interpretation of Fluvial Landforms

Satellite imagery can provide valuable information for interpreting fluvial landforms, which are landforms created by the action of flowing water. Here are some key features to look for in satellite images:

1. Meandering Rivers

Meandering rivers have a sinuous or winding shape, and they often form as a result of erosion and sediment deposition. These can be identified on satellite images by their serpentine shape and the presence of point bars on the inside of meander bends.

India is home to several meandering rivers, which are formed by the erosion and deposition of sediment as the river flows through its channel. Some of the notable meandering rivers in India include:

Ganges River: The Ganges is one of the most famous and important rivers in India, and it meanders through several states, including Uttarakhand, Uttar Pradesh, Bihar, and West Bengal. The river is considered sacred by Hindus and is an important source of water for millions of people.

Brahmaputra River: The Brahmaputra is a major river in Northeast India that meanders through Assam and Arunachal Pradesh before flowing into Bangladesh. The river is known for its rich biodiversity and is home to several endangered species.

Godavari River: The Godavari is a large river that meanders through several states in southern India, including Maharashtra, Telangana, and Andhra Pradesh. The river is an important source of water for agriculture and has several dams and reservoirs built along its course.

Yamuna River: The Yamuna is a major river that meanders through several states in northern India, including Uttarakhand, Uttar Pradesh, Haryana, and Delhi. The river is a tributary of the Ganges and is an important source of water for the region.

These meandering rivers are not only important from an ecological and economic perspective but also serve as popular tourist destinations for river rafting and other adventure activities.

2. Braided Rivers

Braided rivers have multiple channels that divide and recombine, and are typically found in areas with high sediment loads. These can be identified on satellite images by their complex channel patterns, with numerous smaller channels braiding around larger islands or bars.

Braided rivers are a type of river system that are characterized by a network of interconnected channels that are separated by sandbars and islands. These types of rivers are commonly found in areas with high sediment loads and variable flow rates. In India, some of the notable braided rivers are:

Indus River: The Indus is a major river in northern India that flows through the states of Jammu and Kashmir, Himachal Pradesh, and Punjab. The river has a braided morphology in its upper reaches and is an important source of water for irrigation and hydroelectric power generation.

Narmada River: The Narmada is a major river that flows through the states of Madhya Pradesh, Maharashtra, and Gujarat. The river has a braided morphology in its upper reaches and is an important source of water for agriculture and drinking water supply.

Tapti River: The Tapti is a major river that flows through the states of Madhya Pradesh, Maharashtra, and Gujarat. The river has a braided morphology in its upper reaches and is an important source of water for agriculture and industry.

Mahanadi River: The Mahanadi is a major river that flows through the states of Chhattisgarh and Odisha. The river has a braided morphology in its upper reaches and is an important source of water for irrigation, hydroelectric power generation, and industrial use.

These braided river systems are important from an ecological and economic perspective as they provide water for irrigation, hydroelectric power generation, and support a wide range of aquatic and terrestrial wildlife. They also attract tourists interested in adventure activities such as rafting and kayaking.

3. River Terraces

River terraces are flat, elevated surfaces that are created when a river cuts down into its own bed and leaves behind a remnant floodplain. These can be identified on satellite images as flat, raised areas adjacent to a river channel.

River terraces are flat, elevated landforms that are formed by the lateral erosion of rivers over long periods of time. These terraces are often found in areas where the river has cut into the surrounding landscape, leaving behind elevated levels of land along its course. In India, there are several notable river terraces, including:

Tapti River Terraces: The Tapti River, which flows through the states of Maharashtra and Gujarat, has several terraces that are formed due to the lateral erosion of the river. These terraces provide fertile land for agriculture and are also home to several indigenous communities.

Yamuna River Terraces: The Yamuna River, which flows through the states of Uttarakhand, Uttar Pradesh, Haryana, and Delhi, has several terraces that are formed due to the river's lateral erosion. These terraces provide fertile land for agriculture and are also home to several important historical and cultural sites.

Godavari River Terraces: The Godavari River, which flows through the states of Maharashtra, Telangana, and Andhra Pradesh, has several terraces that are formed due to the river's lateral erosion. These terraces provide fertile land for agriculture and are also home to several important pilgrimage sites.

Ganga River Terraces: The Ganga River, which flows through several states in northern India, including Uttarakhand, Uttar Pradesh, Bihar, and West Bengal, has several terraces that are formed due to the river's lateral erosion.

These terraces provide fertile land for agriculture and are also home to several important pilgrimage sites.

These river terraces are important from an ecological and economic perspective as they provide fertile land for agriculture and support a wide range of plant and animal life. They are also important from a cultural and historical perspective as they are home to several important historical and cultural sites.

4. Oxbow Lakes

Oxbow lakes are crescent-shaped bodies of water that form when a meandering river cuts off a meander bend. These can be identified on satellite images as circular or crescent-shaped bodies of water adjacent to a river channel.

Oxbow lakes are a type of landform that are formed by the meandering of rivers. They are essentially U-shaped or horseshoe-shaped bodies of water that are separated from the main river channel. In India, there are several oxbow lakes that are of ecological and cultural significance, including:

Dal Lake: Dal Lake is located in Srinagar, Jammu and Kashmir, and is one of the most famous oxbow lakes in India. The lake is an important source of fish and aquatic plants, and is also a popular tourist destination.

Loktak Lake: Loktak Lake is located in Manipur and is the largest freshwater lake in Northeast India. The lake is a unique ecosystem that supports a wide range of plant and animal life, including several endangered species.

Pulicat Lake: Pulicat Lake is located in Andhra Pradesh and Tamil Nadu and is the second largest brackish water lake in India. The lake is an important breeding ground for several species of migratory birds.

Kanwar Lake: Kanwar Lake is located in Bihar and is the largest freshwater oxbow lake in Asia. The lake is an important ecosystem that supports a wide range of plant and animal life, including several endangered species.

These oxbow lakes are important from an ecological and cultural perspective as they support a wide range of plant and animal life, and are also important sources of livelihood for local communities. They are also popular tourist destinations and attract visitors from all over the world.

5. Delta

Deltas are formed where a river meets a larger body of water, such as a lake or sea, and deposits sediment. They can be identified on satellite images as fan-shaped or triangular landforms at the mouth of a river.

When interpreting satellite images of deltas in India or any other region, there are several key features to consider:

River Channels: Deltas form at the mouths of rivers, so look for the main river channels that transport sediment into the delta. These channels can appear as darker, deeper areas in the image.

Distributary Channels: Deltas often have numerous distributary channels branching out from the main river channels. These channels carry sediment and water across the delta plain. They can appear as lighter or narrower channels in the satellite image.

Sediment Deposition: Deltas are characterized by the deposition of sediment carried by rivers. Look for lighter-colored areas where sediment has accumulated, typically forming flat or gently sloping surfaces.

Vegetation: Deltas usually support dense vegetation due to the fertile soils deposited by rivers. Look for areas with vibrant green colors that indicate vegetation cover. Mangroves are also common in coastal deltas.

Water Bodies: Deltas often have various types of water bodies, such as lakes, lagoons, and marshes. These water bodies can appear as darker areas in the satellite image.

Human Settlements: Deltas are often densely populated areas due to their agricultural productivity and access to water resources. Look for clusters of buildings, roads, and other signs of human settlements.

Remember that interpreting satellite images requires a combination of visual analysis and contextual knowledge of the specific delta region you are studying. It is always helpful to refer to additional resources, such as maps, geological surveys, and local expertise, to gain a deeper understanding of the delta landscape in India.

6. Alluvial Fans

Alluvial fans are cone-shaped landforms that form where a river flows out of a narrow canyon onto a broader plain. They can be identified on satellite images as a wide, flat area at the base of a mountain range, with a fan-shaped pattern of channels and sediment deposits.

Alluvial fans are fan-shaped landforms that are created by the deposition of sediment carried by rivers as they emerge from mountainous regions onto flat or gently sloping plains. Interpreting satellite images of alluvial fans in India involves considering the following features:

Fan Shape: Alluvial fans typically have a fan-shaped appearance, with a broad apex at the base of a mountain range and a tapering shape as they extend away from the mountains. Look for the distinctive fan-like pattern in the satellite image.

River Channels: Alluvial fans are formed by rivers that transport sediment from the mountains. Identify the main river channels that flow from the mountains onto the fan. These channels can appear as darker or more defined features.

Sediment Deposition: Alluvial fans are characterized by the deposition of sediment as the river's energy decreases upon reaching the flatter terrain. Look for areas of lighter-colored sediment deposition, which can appear as flat or gently sloping surfaces.

Channels and Braiding: Alluvial fans may have multiple channels that spread out and braid across the fan surface. These channels can appear as lighter or narrower features, and their patterns can indicate the dynamic nature of the fan system.

Vegetation: Alluvial fans can support vegetation growth, particularly along the river channels and areas with higher water availability. Look for patches of green vegetation that indicate areas of plant cover on the fan.

Human Settlements: Alluvial fans may also have human settlements, as they often provide fertile soils and water resources for agriculture. Look for clusters of buildings, roads, or agricultural fields on or near the fan.

When interpreting satellite images of alluvial fans in India, it's important to consider the regional context and geological information to enhance your understanding. Consulting maps, geological surveys, and local expertise can provide valuable insights into the specific characteristics and dynamics of the alluvial fan you are studying.

By studying these and other features visible on satellite imagery, researchers and geologists can gain insights into the behavior of rivers, the geology of a region, and the environmental conditions that have shaped the landscape over time.

Satellite Images Interpretation of Coastal Landforms

Satellite imagery can also provide valuable information for interpreting coastal landforms, which are landforms that form at the interface between land and sea. Here are some key features to look for in satellite images:

1. Beaches: Beaches are areas of sand or gravel that form at the interface between land and sea. These can be identified on satellite images by their light color and the presence of waves breaking on the shore.
2. Cliffs: Cliffs are steep, vertical rock faces that form as a result of erosion by waves and other coastal processes. These can be identified on satellite images by their dark color and vertical orientation.

3. Dunes: Dunes are mounds of sand that form along the coast as a result of wind action. These can be identified on satellite images by their light color and distinctive crescent-shaped pattern.
4. Barrier Islands: Barrier islands are long, narrow islands that form parallel to the coast and provide protection for the mainland from storms and waves. These can be identified on satellite images by their long, narrow shape and the presence of a lagoon or estuary on the landward side.
5. Estuaries: Estuaries are coastal bodies of water where freshwater from rivers mixes with saltwater from the ocean. These can be identified on satellite images by their dark color and the presence of sinuous channels where the river water enters the estuary.
6. Coral Reefs: Coral reefs are underwater structures that form as a result of the accumulation of coral skeletons. These can be identified on satellite images by their distinctive bright color and circular or semicircular shape.

By studying these and other features visible on satellite imagery, researchers and coastal managers can gain insights into the behavior of the coast, the geology of a region, and the environmental conditions that have shaped the landscape over time. This information can be used to inform coastal management decisions, such as where to build infrastructure and how to manage coastal hazards.

Important Questions

1. What factors influence the formation of different drainage patterns?
2. How can the study of drainage patterns help in understanding the geological history of an area?
3. What are the main types of drainage patterns observed in river systems?
4. How do dendritic and trellis drainage patterns differ in terms of their formation and characteristics?
5. What are the implications of a parallel drainage pattern in a landscape?
6. How does a radial drainage pattern indicate the presence of a volcanic or domal feature?
7. Can the analysis of drainage patterns be used to identify area prone to flooding or landslides?
8. How do rectangular and angular drainage patterns develop in areas with jointed or faulted bedrock?

9. What are the potential environment implications of disrupted drainage patterns due to human activities?
10. How can modern technologies like remote sensing and GIS aid in the interpretation and analysis of drainage patterns?

5

Interpretation of Structural and Denudational Landforms

Structural landforms are landforms that are created due to the underlying geological structures of the earth's crust. These structures can include folds, faults, and other features that have been formed due to the tectonic forces that have shaped the earth's surface over millions of years.

Examples of structural landforms include mountain ranges, plateaus, and rift valleys. Mountain ranges are formed due to the upliftment of the earth's crust along a fault line or due to the folding of the earth's crust. Plateaus are flat-topped landforms that are formed due to the upliftment of a large area of the earth's crust. Rift valleys are formed due to the stretching and pulling apart of the earth's crust along a fault line.

Denudational landforms, on the other hand, are landforms that are created due to the process of erosion and weathering. These landforms are formed due to the gradual wearing down of the earth's surface by natural forces such as wind, water, and ice.

Examples of denudational landforms include valleys, canyons, and cliffs. Valleys are formed due to the erosive action of rivers, which carve out channels over time. Canyons are deep, narrow valleys that are usually formed due to the erosion of rock by water. Cliffs are vertical or near-vertical walls of rock that are formed due to the erosive action of waves, wind, or water.

In summary, structural landforms are formed due to the underlying geological structures of the earth's crust, while denudational landforms are formed due to the process of erosion and weathering. Both types of landforms play a critical role in shaping the earth's surface and providing habitats for a wide variety of plant and animal species.

Cuesta

Cuesta is a geological term used to describe a type of landform that consists of a ridge or a hill with a gentle slope on one side and a steep slope on the other side. The gentle slope is called the dip slope or the backslope, while the steep slope is called the escarpment or the face slope.

Cuestas are formed due to the erosion of sedimentary rocks, which have alternating layers of hard and soft rock. The hard rock layers are more resistant to erosion than the soft rock layers, and as a result, the soft rock layers are eroded more quickly. This differential erosion results in the formation of a ridge with a gentle slope on one side and a steep slope on the other side.

Cuestas are commonly found in areas with gently rolling terrain, such as the Great Plains in the United States, where they often form the boundary between different types of geological formations. Cuestas are also found in other parts of the world, including Europe, Africa, and Australia.

Cuestas can have important economic and ecological significance. The gently sloping backslope of a cuesta is often used for agriculture, while the steep escarpment can provide habitat for unique plant and animal species. The cuesta's topography can also influence the flow of water and sediment in rivers and streams, which can impact the ecology of downstream ecosystems.

Hogback

A hogback is a narrow, sharp-crested ridge that is formed by the erosion of steeply dipping rock layers. Hogbacks are typically found in areas with sedimentary rock formations, where the layers of rock have been tilted or folded due to tectonic forces.

Hogbacks are usually several meters to tens of meters high, with a steep slope on one side and a gentler slope on the other. They are often long and sinuous, following the trend of the underlying rock layers.

Hogbacks are formed due to differential erosion of the steeply dipping rock layers. The more resistant layers of rock are eroded more slowly, leaving behind a ridge with a sharp crest. The gentler slope on the other side of the ridge is formed by the erosion of the softer, more easily eroded layers of rock.

Hogbacks are commonly found in arid and semi-arid regions, where erosion rates are high due to the lack of vegetation cover and intense rainfall events. They can also be found in mountainous regions, where they may form the boundary between different rock formations.

Hogbacks can have important ecological and economic significance. They can provide habitat for unique plant and animal species, and their topography can influence the flow of water and sediment in rivers and streams. They may also be used as a source of building stone or as a natural barrier for infrastructure development.

Butte

A butte is a flat-topped hill or mountain with steep sides and a small, relatively flat summit area. Buttes are typically found in arid and semi-arid regions where erosion rates are high and the landscape is characterized by flat-lying sedimentary rock formations.

Buttes are formed by a combination of erosion and tectonic uplift. The process starts with the deposition of sedimentary rocks in a basin or valley. These rocks are then uplifted by tectonic forces, exposing them to the erosive action of wind and water. Over time, the softer, less resistant rocks are eroded away, leaving behind a steep-sided, flat-topped hill or mountain.

The flat top of a butte is typically made up of a layer of hard, resistant rock, which has been able to withstand erosion better than the surrounding rocks. The steep sides of the butte are formed by the erosion of softer, less resistant rocks that once surrounded the hard layer.

Buttes are often found in groups or clusters, and they may be associated with other landforms such as mesas, plateaus, and canyons. The formation of buttes is influenced by a number of factors, including the type and structure of the underlying rocks, the climate and precipitation patterns, and the rate of erosion.

Buttes can have important ecological and cultural significance. They provide habitat for a wide range of plant and animal species, and they may be considered sacred or important cultural sites by indigenous communities. They are also popular tourist attractions, drawing visitors to their dramatic, otherworldly landscapes.

Mesa

A mesa is a flat-topped mountain or hill with steep sides, usually found in arid or semi-arid regions. Mesas are similar to buttes but larger in size, with a broader flat top that can span several square miles.

Mesas are formed by a combination of geological processes including sediment deposition, uplift, and erosion. Layers of sedimentary rock are deposited in a basin or valley, and then are uplifted by tectonic forces. The uplifted rocks are then subjected to erosion by wind and water, which removes the softer, less resistant rocks and leaves behind a flat-topped mesa with steep sides.

The flat top of a mesa is usually composed of a layer of hard, resistant rock, such as sandstone or limestone, which is more resistant to erosion than the surrounding rock layers. The steep sides of a mesa are formed by the erosion of softer, less resistant rock layers.

Mesas are often found in clusters or groups, and may be associated with other landforms such as canyons, plateaus, and buttes. They are important features of the landscape in many arid and semi-arid regions, and are home to a variety of plant and animal species that are adapted to life in these harsh environments.

Mesas can also have cultural and historical significance. They have been used as landmarks and gathering places by indigenous peoples for centuries, and many contain important archaeological sites that provide insights into past human cultures and civilizations. Today, mesas are popular tourist attractions, drawing visitors to their scenic vistas and unique geological features.

Structural and Denudational Landforms in India

India is characterized by a diverse range of landforms that have been shaped by various geological processes over millions of years. Two prominent categories of landforms in India are structural landforms and denudational landforms.

Structural Landforms

Structural landforms are created by the deformation and movement of the Earth's crust. In India, some significant structural landforms include:

Himalayas: The Himalayan mountain range, stretching across northern India, is one of the most prominent structural landforms. It is formed due to the collision between the Indian and Eurasian tectonic plates. The Himalayas encompass towering peaks, deep valleys, and glacier-carved landscapes.

Western Ghats and Eastern Ghats: These mountain ranges run parallel to the western and eastern coasts of India, respectively. They are also a result of tectonic activity. The Western Ghats are known for their lush forests, waterfalls, and high plateaus, while the Eastern Ghats comprise hills, plateaus, and deep valleys.

Aravalli Range: Located in western India, the Aravalli Range is one of the oldest fold mountain systems in the world. It stretches across the states of Rajasthan, Haryana, and Gujarat. The Aravalli Range features rocky outcrops, forested hills, and several important mineral deposits.

Denudational Landforms

Denudational landforms are shaped by the processes of weathering, erosion, and deposition. In India, several denudational landforms can be observed:

Plateaus: Plateaus are large elevated flat areas characterized by steep slopes and cliffs. The Deccan Plateau in central and southern India is a notable example. It was formed by volcanic activity and subsequent weathering and erosion.

Valleys: Various types of valleys can be found across India. For example, the Kashmir Valley in Jammu and Kashmir is a classic example of a glacial valley formed by the action of glaciers. The Narmada Valley in central India is a rift valley formed by tectonic activity.

Alluvial Plains: Alluvial plains are formed by the deposition of sediments carried by rivers. The Indo-Gangetic Plain, stretching across northern India, is the largest alluvial plain in the country. It is characterized by fertile soil and is known for its agricultural productivity.

Coastal Landforms: India has a vast coastline with diverse coastal landforms. These include sandy beaches, dunes, estuaries, lagoons, and deltas. Examples include the Rann of Kutch in Gujarat, the Sundarbans delta in West Bengal, and the beaches of Goa and Kerala.

These are just a few examples of the structural and denudational landforms found in India. The country's geological diversity has shaped its landscapes and contributes to its natural beauty and ecological significance.

Applications of Remote Sensing and GIS

Remote sensing and Geographic Information System (GIS) technologies play a crucial role in studying structural and denudational landforms by providing valuable spatial data and analytical tools. Here are some applications of remote sensing and GIS in the study of structural and denudational landforms:

Landform Mapping: Remote sensing data, including satellite imagery and aerial photographs, can be used to identify and map different landforms. This helps in understanding the distribution, extent, and characteristics of structural and denudational landforms over large areas. GIS software enables the integration and analysis of this spatial data, allowing researchers to create accurate landform maps.

Terrain Analysis: Remote sensing and GIS can be utilized to perform terrain analysis, which involves studying landforms' morphological characteristics and their relationship with other geographic features. Elevation data derived from remote sensing sources, such as digital elevation models (DEMs), are used to analyze slope, aspect, curvature, and other terrain attributes. This aids in identifying structural landforms like mountain ranges, ridges, and valleys.

Change Detection: Remote sensing data collected over different time periods can be compared to detect changes in landforms. By analyzing multispectral or radar imagery, researchers can identify shifts in landform boundaries, erosion or deposition patterns, and other changes caused by natural processes or human activities. GIS helps in quantifying and visualizing these changes over time.

Geomorphic Analysis: Remote sensing and GIS techniques can be employed to conduct geomorphic analysis, which involves studying landforms' evolution and processes. By integrating remote sensing data with topographic and hydrological information, researchers can analyze erosion, sedimentation, and fluvial dynamics. This aids in understanding denudational landforms like valleys, river systems, and coastal features.

Landform Classification and Modeling: Remote sensing data, combined with GIS, can support landform classification and modeling. By extracting relevant spectral, textural, and morphological parameters from remote sensing imagery, researchers can develop algorithms and models to classify landforms automatically. This enables the identification and characterization of specific structural and denudational landforms at regional or global scales.

Environmental Monitoring: Remote sensing and GIS are valuable tools for monitoring and assessing the impacts of natural hazards on landforms. For instance, monitoring techniques using satellite imagery can help detect landslides, coastal erosion, or changes in river channels. GIS enables the integration of various environmental data, facilitating comprehensive hazard assessment and risk management.

By utilizing remote sensing and GIS technologies, researchers can obtain a comprehensive understanding of structural and denudational landforms. These tools enable accurate mapping, analysis of morphological characteristics, change detection, modeling, and environmental monitoring, contributing to improved landform studies and land management practices.

Important Questions

1. How do structural landforms differ from denudational landforms in terms of their formation processes?
2. What are some examples of structural landforms and how are they formed?
3. What factors contribute to the development of denudational landforms?
4. How can the study of structural landforms help in understanding the geological history of a region?
5. What are the main characteristics of denudational landforms and how do they vary across different landscapes?
6. What are the major agents of erosion and depositional responsible for the information of denudational landforms?
7. How does climate influence the development and preservation of structural landforms?
8. What are the distinguishing features of fluvial landforms and how are they related to denudational processes?
9. How do tectonic activities contribute to the formation of structural landforms?
10. What role does time play in the formation and evolution of both structural and denudational landforms?

6

Interpretation of Landforms Related to Different Rock Types

Rock

A rock is a solid, naturally occurring substance composed of minerals or mineraloids. It is typically formed through geological processes, such as cooling and solidification of molten lava or magma, or by the deposition and compaction of sediment over time. Rocks are classified into three main types based on their origin: igneous, sedimentary, and metamorphic.

Igneous rocks: Formed from the cooling and solidification of molten lava or magma. Examples include granite, basalt, and obsidian.

Sedimentary rocks: Formed from the accumulation and compaction of sediment, such as sand, mud, or organic matter, over long periods. Examples include sandstone, limestone, and shale.

Metamorphic rocks: Formed from the transformation of existing rocks under high temperature and pressure conditions. They undergo changes in mineral composition, texture, and structure. Examples include marble, slate, and gneiss.

Rocks serve as the building blocks of the Earth's crust and are an essential part of the geological processes that shape the planet. They can provide valuable information about Earth's history, including the formation of mountains, the movement of continents, and the presence of ancient life forms. Rocks are also used in various practical applications, such as construction, road building, and the production of minerals and resources.

Rock Cycle

The rock cycle is a continuous process that describes the formation, transformation, and recycling of rocks in the Earth's crust. It involves the interplay between the three main types of rocks: igneous, sedimentary, and metamorphic. The rock cycle can be summarized in the following steps:

Igneous Rock Formation: The rock cycle begins with the formation of igneous rocks. Igneous rocks are formed when molten lava or magma cools and solidifies. This process can occur either beneath the Earth's surface (intrusive igneous rocks) or on the surface (extrusive igneous rocks).

Weathering and Erosion: Over time, igneous rocks are subjected to weathering and erosion, which break them down into smaller fragments and sediments. Weathering can occur due to physical, chemical, and biological processes, such as wind, water, temperature changes, and the action of plants and organisms.

Sediment Deposition: The weathered fragments and sediments are transported by wind, water, or ice and are eventually deposited in riverbeds, lakes, oceans, or other sedimentary environments. Over time, these sediments accumulate and undergo compaction and cementation to form sedimentary rocks.

Lithification: The process of lithification involves the compaction and cementation of sedimentary particles under pressure. This transforms the loose sediments into solid sedimentary rocks, such as sandstone, limestone, or shale.

Metamorphism: If sedimentary or igneous rocks are subjected to high temperature and pressure, they can undergo metamorphism. During metamorphism, the rocks undergo changes in mineral composition, texture, and structure, resulting in the formation of metamorphic rocks, such as marble, slate, or gneiss.

Melting and Magma Formation: When rocks experience intense heat and pressure in the Earth's interior, they can melt and become molten magma. This can occur in subduction zones, where tectonic plates collide, or in areas of intense volcanic activity. The molten magma can then rise to the surface and solidify as igneous rocks, continuing the cycle.

The rock cycle is a continuous and dynamic process that occurs over millions of years. It highlights the interconnectedness and transformation of rocks through geological processes, driven by forces such as plate tectonics, erosion, and thermal activity. The rock cycle plays a crucial role in shaping the Earth's surface, creating geological features, and providing insights into the Earth's history and composition.

Major Rock Types in India

India has a diverse geological landscape, and as a result, various rock types can be found throughout the country. Here are some of the major rock types in India:

Archean Rocks: These are some of the oldest rocks in India, dating back billions of years. Archean rocks include granite, gneiss, and schist, and they can be found in regions such as the Dharwar craton in Karnataka and the Singhbhum craton in Jharkhand.

Deccan Basalt: The Deccan Traps, located in western and central India, are composed of extensive basaltic lava flows. These volcanic rocks were formed around 65 million years ago during the Cretaceous period and cover large parts of Maharashtra, Madhya Pradesh, Gujarat, and Telangana.

Sedimentary Rocks: Sedimentary rocks are widespread throughout India and are found in various geological formations. Some notable sedimentary rock types include sandstone, limestone, shale, and conglomerate. The Vindhyan Range in central India and the Siwalik Hills in the northern part of the country are known for their sedimentary rock formations.

Gondwana Rocks: Gondwana rocks are associated with the Gondwana supercontinent and were formed during the Permian and Triassic periods. These rocks are primarily sedimentary in nature and consist of sandstones, shales, and coal deposits. They can be found in regions like Jharkhand, Odisha, Chhattisgarh, and Madhya Pradesh.

Metamorphic Rocks: Metamorphic rocks are present in various parts of India and have undergone changes in mineral composition and structure due to high temperature and pressure. They include rocks such as marble, quartzite, slate, and schist. Metamorphic terrains can be found in regions like the Aravalli Range in Rajasthan, the Eastern Ghats, and parts of the Himalayas.

Granitic Rocks: Granitic rocks, including granite and granodiorite, are widespread in India. They are associated with plutonic intrusions and can be found in regions like the Aravalli Range, Eastern Ghats, Western Ghats, and the Himalayas.

These are just a few examples of the major rock types in India. The geological diversity of the country is vast, and different regions have their own unique rock formations, influenced by factors such as tectonic activity, geological history, and local environmental conditions.

Interpretation of Landforms Related to Igneous Rock

Igneous rocks are formed from the solidification and crystallization of magma or lava. The landforms related to igneous rocks are often a result of volcanic activity, which can be classified into two types: extrusive and intrusive.

Extrusive landforms related to igneous rocks are formed when lava flows out of a volcano and cools and solidifies on the surface. These landforms include volcanic cones, lava fields, and lava plateaus. Volcanic cones can be steep-sided or gently sloping, and they are formed when lava is ejected from a central vent and builds up around it. Lava fields are extensive areas covered by solidified lava flows, and they can be smooth or rough depending on the viscosity of the lava. Lava plateaus are flat-topped mountains or high plateaus made up of layers of solidified lava.

Intrusive landforms related to igneous rocks are formed when magma solidifies beneath the surface of the Earth. These landforms include batholiths, dikes, and sills. Batholiths are large masses of intrusive igneous rock that have crystallized deep beneath the Earth's surface. Dikes are vertical or near-vertical sheets of igneous rock that cut through the surrounding rock, while sills are horizontal sheets of igneous rock that are parallel to the surrounding rock.

Other landforms related to igneous rocks include volcanic necks, which are the solidified remains of the magma conduit that once fed a volcano, and volcanic plugs, which are solidified magma that has solidified within the conduit and effectively blocked it.

In summary, landforms related to igneous rocks are diverse and can be either extrusive or intrusive. Extrusive landforms include volcanic cones, lava fields, and lava plateaus, while intrusive landforms include batholiths, dikes, and sills. Other landforms related to igneous rocks include volcanic necks and volcanic plugs.

Interpretation of Landforms Related to Sedimentary Rocks

Sedimentary rocks are formed by the accumulation of sediment or organic matter over time. The landforms related to sedimentary rocks are typically shaped by the processes of erosion, deposition, and compaction that occur during the formation of sedimentary rocks.

One of the most common landforms related to sedimentary rocks is a sedimentary basin, which is a low-lying area where sediment has accumulated over time. Sedimentary basins can be caused by a variety of factors, including tectonic activity, erosion, and sea-level changes. Examples of sedimentary basins include the Gulf of Mexico and the Great Basin of the western United States.

Another common landform related to sedimentary rocks is a sedimentary outcrop, which is an exposed section of sedimentary rock that can be used to study the rock's characteristics and history. Sedimentary outcrops can be found in a variety of settings, including cliffs, canyons, and road cuts.

Sedimentary rocks can also form distinctive landforms such as limestone karsts, which are characterized by underground drainage systems with sinkholes and caves. Limestone karsts are formed when acidic water dissolves the rock over time, resulting in unique and often striking topography.

Other landforms related to sedimentary rocks include sand dunes, which are formed by windblown sand particles, and river valleys, which are shaped by the erosive action of flowing water and the deposition of sediment. Coastal cliffs and beaches are also shaped by the deposition of sediment, as well as the action of waves and tides.

In summary, landforms related to sedimentary rocks are shaped by the processes of erosion, deposition, and compaction that occur during the formation of sedimentary rocks. Common landforms include sedimentary basins, sedimentary outcrops, limestone karsts, sand dunes, river valleys, and coastal cliffs and beaches.

Interpretation of Landforms Related to Metamorphic Rocks

Metamorphic rocks are formed from pre-existing rocks that have been subjected to intense heat, pressure, or chemical activity. The landforms related to metamorphic rocks are typically characterized by rugged and complex topography, reflecting the dynamic geological processes that have shaped them.

One of the most common landforms related to metamorphic rocks is a mountain range, which is typically formed by the collision of tectonic plates and the subsequent folding and faulting of rocks. The rocks in mountain ranges are often metamorphic, as they have been subjected to intense heat and pressure during the tectonic processes that formed the mountains.

Another landform related to metamorphic rocks is a metamorphic dome, which is a circular or elongated structure that is formed by the upward displacement of rocks due to intense pressure from below. Metamorphic domes are often associated with the core of mountain ranges and are typically composed of high-grade metamorphic rocks, such as gneiss.

Slate quarries are also a distinctive landform related to metamorphic rocks. Slate is a fine-grained metamorphic rock that is easily split into thin, flat sheets, and is often used as roofing tiles or as a writing surface. Slate quarries are typically found in hilly or mountainous areas where the rock has been subjected to intense pressure and has undergone metamorphism.

Other landforms related to metamorphic rocks include fault scarps, which are steep slopes that are formed by the displacement of rocks along a fault line, and metamorphic outcrops, which are exposed sections of metamorphic rock that can be studied to learn about the rock's history and characteristics.

In summary, landforms related to metamorphic rocks are characterized by rugged and complex topography, reflecting the intense geological processes that have shaped them. Common landforms include mountain ranges, metamorphic domes, slate quarries, fault scarps, and metamorphic outcrops.

Important Questions

1. How does the presence of limestone affect the formation of karst landscape?
2. What landforms are commonly associated with granite rock types?
3. How does the erosion of sandstone differ from that of shale?
4. What types of landforms are typically found in areas with basalt rock formations?
5. How do sedimentary rock formations contribute to the development of river valley?
6. What landforms are commonly associated with volcanic rock types?
7. How does the weathering of granite differ from that of marble?
8. What landforms are typically formed by the erosion of coastal cliffs composed of sandstone?
9. How does the presence of quartzite affect the information of desert landscapes?
10. What landforms are commonly associated with metamorphic rock types?

7

Geomorphological Mapping and Terrain Evaluation

Geomorphological mapping is the process of creating detailed maps that illustrate the physical features of the Earth's surface, including landforms, soil types, vegetation, and water resources. These maps can be used to better understand the underlying geology and topography of a region, and can be used for a variety of purposes, such as land-use planning, environmental management, and geological hazard assessment.

Geomorphological maps are typically created using a combination of field observations, aerial photography, and remote sensing data. Field observations involve physically visiting a site and making detailed observations of the landforms, soil types, and vegetation. Aerial photography involves taking high-resolution photographs of the Earth's surface from an airplane or satellite, which can be used to create detailed maps of the area.

Remote sensing data can also be used to create geomorphological maps, including data from satellites, radar, and LiDAR (Light Detection and Ranging) systems. These technologies allow scientists to gather data about the Earth's surface without physically visiting the site, and can provide highly accurate information about the topography, geology, and vegetation of an area.

The resulting maps can be used for a variety of purposes, including land-use planning, natural resource management, and hazard assessment. For example, geomorphological maps can be used to identify areas of high risk for landslides, floods, or earthquakes, and can be used to develop strategies for managing these hazards. They can also be used to identify areas of high conservation value, such as wetlands or other ecologically sensitive areas, and can be used to develop management plans that protect these resources.

In summary, geomorphological mapping is an important tool for understanding the physical features of the Earth's surface and can be used for a variety of purposes, including land-use planning, environmental management, and geological hazard assessment. The process involves a combination of field observations, aerial photography, and remote sensing data to create detailed maps of the landforms, soil types, and vegetation of a region.

Geomorphological Mapping of River in India

India is a country with a vast network of rivers, including major ones such as the Ganges, Brahmaputra, and Indus. Geomorphological mapping of these rivers is an important tool for understanding their characteristics, including their flow patterns, sediment transport, and flood hazards.

The process of geomorphological mapping of rivers in India typically involves a combination of field observations, remote sensing data, and numerical modeling. Field observations involve physically visiting the river and making detailed observations of the channel morphology, sediment types, and vegetation. Remote sensing data, including satellite imagery and LiDAR data, can provide additional information about the river's topography, as well as its floodplain and surrounding terrain.

Numerical modeling can be used to simulate the river's hydrodynamics, sediment transport, and flood behavior. This can help to identify areas of high erosion or deposition, as well as areas of high flood risk.

The resulting geomorphological maps can be used for a variety of purposes, including river management, flood control, and environmental protection. For example, they can be used to identify areas where erosion is occurring, which can help to develop strategies to prevent further erosion and maintain the stability of the river channel. They can also be used to identify areas of high flood risk, which can help to develop strategies for flood control and disaster management.

Geomorphological mapping of rivers in India is an important tool for understanding their characteristics and managing their resources in a sustainable and effective manner.

Geomorphological Mapping of Coastal Area in India

Geomorphological mapping of coastal areas in India is an important tool for understanding the characteristics of the coastline, including the landforms, sediment types, and coastal processes that shape the shoreline. India has a vast coastline of over 7,500 km, including the coasts along the Arabian Sea, Bay of Bengal, and the Indian Ocean.

The process of geomorphological mapping of coastal areas in India typically involves a combination of field observations, remote sensing data, and numerical modeling. Field observations involve physically visiting the coastline and making detailed observations of the landforms, sediment types, and vegetation. Remote sensing data, including satellite imagery, aerial photography, and LiDAR data, can provide additional information about the coastline, including topography, coastal features, and vegetation cover.

Numerical modeling can be used to simulate coastal processes, such as wave and tidal action, sediment transport, and erosion and deposition patterns. This can help to identify areas of high erosion or deposition, as well as areas of high flood or coastal hazard risk.

The resulting geomorphological maps can be used for a variety of purposes, including coastal zone management, coastal erosion control, and disaster management. They can help to identify areas where erosion is occurring and to develop strategies to prevent further erosion, such as the construction of coastal structures like breakwaters and seawalls. They can also help to identify areas of high flood or coastal hazard risk, which can help to develop strategies for disaster management and mitigation.

Geomorphological mapping of coastal areas in India is an important tool for understanding the characteristics of the coastline, managing its resources in a sustainable manner, and mitigating the impact of coastal hazards on the population and the environment.

Geomorphological Mapping of Glacial Area in India

Geomorphological mapping of glacial areas in India is an important tool for understanding the characteristics of the glaciers and the surrounding terrain, including landforms, sediment types, and glacial processes that shape the landscape. India has several important glacier regions, including the Himalayas, Karakoram, and the Trans-Himalayas.

The process of geomorphological mapping of glacial areas in India typically involves a combination of field observations, remote sensing data, and numerical modeling. Field observations involve physically visiting the glaciers and the surrounding terrain and making detailed observations of the landforms, sediment types, and vegetation. Remote sensing data, including satellite imagery, aerial photography, and LiDAR data, can provide additional information about the topography, glacier extent, and glacial features.

Numerical modeling can be used to simulate glacial processes, such as ice flow, meltwater runoff, and glacier retreat. This can help to identify areas of high glacial retreat or deposition, as well as areas of high glacial hazard risk.

The resulting geomorphological maps can be used for a variety of purposes, including glacier monitoring, hazard assessment, and climate change research. They can help to identify areas where glacier retreat is occurring and to develop strategies to mitigate the impacts of climate change, such as water resource management and disaster management planning. They can also help to identify areas of high glacial hazard risk, which can help to develop strategies for disaster management and mitigation.

Geomorphological mapping of glacial areas in India is an important tool for understanding the characteristics of the glaciers and their impact on the environment, managing their resources in a sustainable manner, and mitigating the impact of glacial hazards on the population and the environment.

Terrain Evaluation Using Satellite Data

Terrain evaluation using satellite data can provide valuable insights into the topography and physical features of the earth's surface. Satellite data can be used to generate accurate digital elevation models (DEMs) which are important for various applications such as hydrology, natural resource management, and urban planning.

Satellite imagery can also be used to identify various terrain features such as mountains, valleys, rivers, and lakes. This information can be used to create detailed maps and 3D models of the terrain. In addition, satellite data can be used to monitor changes in the terrain over time, such as erosion or the growth of vegetation.

One of the most common methods for evaluating terrain using satellite data is through the use of radar. Radar signals can penetrate through clouds and vegetation, allowing for accurate measurements of the terrain. This is particularly useful in areas with dense vegetation or cloud cover, where other forms of remote sensing may not be effective.

Another method for evaluating terrain using satellite data is through the use of LiDAR (Light Detection and Ranging) technology. LiDAR uses laser pulses to generate highly accurate elevation data, which can be used to create detailed 3D models of the terrain. LiDAR can also be used to identify various terrain features, such as trees and buildings, and to generate accurate maps of the terrain.

Terrain evaluation using satellite data is a powerful tool for understanding the earth's surface and can be used for a wide range of applications.

Importance of Satellite Data in Terrain Evaluation

Satellite data plays a crucial role in terrain evaluation as it provides a comprehensive and accurate view of the earth's surface. Here are some of the key reasons why satellite data is important in terrain evaluation:

- Large coverage: Satellite data provides a vast coverage area of the earth's surface, making it possible to evaluate remote and inaccessible areas. This is particularly useful for monitoring terrain features such as mountains, oceans, and polar regions.

- High-resolution imagery: Satellite data can provide high-resolution imagery, which allows for detailed analysis of terrain features. This helps to identify landforms, water bodies, vegetation, and other natural features that are critical for understanding terrain dynamics.
- Consistent monitoring: Satellite data provides a consistent monitoring system for terrain evaluation over time. It enables the detection of changes in terrain, such as erosion, land cover changes, and vegetation growth, which is essential for land management and natural resource conservation.
- Accessibility: Satellite data is easily accessible and can be shared across different organizations and agencies. This enables various stakeholders to access the same information, facilitating better collaboration and coordination in terrain evaluation.
- Cost-effective: Using satellite data for terrain evaluation is often more cost-effective than traditional field-based methods. This is because satellite data can cover large areas in a short period, reducing the need for extensive fieldwork.

In summary, satellite data is crucial for terrain evaluation as it provides accurate and reliable information on the earth's surface, which is essential for a wide range of applications, including natural resource management, environmental monitoring, and disaster management.

Important Questions

1. What are the main objectives of geomorphological mapping?
2. What are the key factors to consider when evaluating terrain for engineering project?
3. What are the different methods and techniques used in geomorphological mapping?
4. How does the understanding of landforms contribute to terrain evaluation?
5. What role does remote sensing play in geomorphological mapping and terrain evaluation?
6. What are the challenges faced in mapping and evaluating terrains in inaccessible areas?
7. How can geomorphological mapping aid in hazard assessment and mitigation?
8. What are the implications of land use changes on terrain evaluation?

9. How can digital elevation model (DEMs) be used in terrain evaluation?
10. What are the applications of geomorphological mapping and terrain evaluation in environmental management?

8

Remote Sensing in Mineral Exploration

Remote Sensing in Mineral Exploration

Remote sensing is a powerful tool used in mineral exploration to study and identify minerals and mineral deposits from a distance. This technology allows geologists to gather information about the Earth's surface and subsurface without having to physically access the site. Here are some ways remote sensing is used in mineral exploration:

Mapping Geology: Remote sensing can be used to map geology by identifying surface features associated with mineral deposits. Geologists use images collected by remote sensing satellites and aircraft to create detailed maps of the geology and topography of a region. These maps can help identify areas that are likely to contain mineral deposits.

Mineral Identification: Remote sensing can help identify specific minerals by analyzing the reflectance or absorption patterns of light. Different minerals have unique spectral signatures that can be detected by remote sensing instruments. By analyzing these signatures, geologists can identify the type and abundance of minerals present in an area.

Target Identification: Remote sensing can be used to identify specific areas that are likely to contain mineral deposits. Geologists can use remote sensing data to identify geological features, such as faults or folds, that are often associated with mineral deposits. They can also look for areas with anomalous mineral signatures, which may indicate the presence of mineral deposits.

Environmental Monitoring: Remote sensing can be used to monitor the environmental impact of mining operations. By comparing images collected before and after mining, geologists can track changes in land use, vegetation cover, and water quality.

Remote sensing is a powerful tool in mineral exploration that can save time and money while providing valuable information about the geology and mineral resources of a region.

Overview of Remote Sensing in Mineral Exploration

Remote sensing is an essential technique used in mineral exploration to gather data about the Earth's surface and subsurface without physically accessing the site. It involves the use of sensors mounted on satellites, airplanes, drones, or ground-based platforms to collect data about the Earth's surface and atmosphere.

In mineral exploration, remote sensing is used to identify and map geological features associated with mineral deposits, detect the presence of minerals, and monitor the environmental impact of mining operations. Here are some of the ways remote sensing is used in mineral exploration:

Mapping geology: Remote sensing can be used to create detailed maps of the geology and topography of an area. By analyzing images collected by remote sensing instruments, geologists can identify surface features associated with mineral deposits and create maps that can help guide exploration activities.

Mineral identification: Remote sensing can detect the spectral signatures of minerals and can be used to identify the type and abundance of minerals present in an area. By analyzing the reflectance or absorption patterns of light, geologists can identify specific minerals and assess their potential economic value.

Target identification: Remote sensing can help identify areas that are likely to contain mineral deposits. Geologists can use remote sensing data to identify geological features, such as faults or folds, that are often associated with mineral deposits. They can also look for areas with anomalous mineral signatures that may indicate the presence of mineral deposits.

Environmental monitoring: Remote sensing can be used to monitor the environmental impact of mining operations. By comparing images collected before and after mining, geologists can track changes in land use, vegetation cover, and water quality.

In summary, remote sensing is a valuable tool in mineral exploration that can provide geologists with critical information about the geology and mineral resources of an area. It can save time and money while improving exploration success rates and minimizing environmental impact.

Application of Remote Sensing in Mineral Exploration

Remote sensing is a widely used technique in mineral exploration due to its ability to collect data about the Earth's surface and subsurface without physically accessing the site. Here are some specific applications of remote sensing in mineral exploration:

Mapping geology: Remote sensing can be used to create detailed maps of the geology and topography of an area. By analyzing images collected by remote sensing instruments, geologists can identify surface features associated with mineral deposits and create maps that can help guide exploration activities.

Mineral identification: Remote sensing can be used to identify the type and abundance of minerals present in an area. By analyzing the reflectance or absorption patterns of light, geologists can identify specific minerals and assess their potential economic value.

Target identification: Remote sensing can help identify areas that are likely to contain mineral deposits. Geologists can use remote sensing data to identify geological features, such as faults or folds, that are often associated with mineral deposits. They can also look for areas with anomalous mineral signatures that may indicate the presence of mineral deposits.

Alteration mapping: Remote sensing can be used to map the distribution of hydrothermal alteration minerals associated with mineral deposits. Hydrothermal alteration minerals are formed by the alteration of rocks by hot fluids associated with mineral deposits. These minerals have characteristic spectral signatures that can be detected by remote sensing instruments.

Structural analysis: Remote sensing can be used to identify and map structural features, such as faults and fractures, that can control the distribution of mineral deposits. By analyzing images collected by remote sensing instruments, geologists can identify structural features and use this information to develop exploration targets.

Remote sensing is a valuable tool in mineral exploration that can provide geologists with critical information about the geology and mineral resources of an area. It can save time and money while improving exploration success rates and minimizing environmental impact.

Important Questions

1. How can remote sensing techniques be used in mineral exploration?
2. What types of remote sensing data are commonly utilized in mineral exploration?
3. How does hyperspectral remote sensing aid in mineral identification and mapping?
4. What are the advantages of using remote sensing for mineral exploration compared to traditional methods?
5. How can remote sensing techniques help in locating hidden or buried mineral deposits?
6. What role does aerial photography play in mineral exploration using remote sensing?
7. What are the limitations or challenges associated with remote sensing in mineral exploration?

8. How can thermal remote sensing be employed in detecting specific mineral deposits?
9. What are some common algorithms or techniques used to process and analyse remote sensing data for mineral exploration purposes?
10. How can remote sensing assist in monitoring changes in mineral resources over time?

9

Remote Sensing in Oil Exploration

Remote Sensing in Oil Exploration

Remote sensing plays a significant role in the oil exploration industry. It refers to the use of sensors and instruments to gather information about the Earth's surface and its environment from a distance. This information can be used to locate and evaluate oil and gas deposits.

There are different remote sensing techniques used in oil exploration, including

Seismic surveys: This technique involves creating sound waves underground and recording their reflection to determine the subsurface structure. It is commonly used in oil exploration to locate potential oil and gas deposits.

Magnetic surveys: This technique uses a magnetometer to measure the magnetic field of the Earth's surface. Magnetic anomalies can indicate the presence of underground geological structures that may contain oil or gas deposits.

Gravity surveys: This technique involves measuring variations in the Earth's gravitational field, which can indicate the presence of underground geological structures.

Remote sensing imagery: This technique uses satellite or aerial imagery to identify surface features that may indicate the presence of oil or gas deposits. For example, vegetation patterns, soil moisture levels, and temperature variations can indicate the presence of hydrocarbons.

Remote sensing data can also be used for monitoring and managing oil and gas production. For example, satellite imagery can be used to monitor pipeline networks and detect leaks. In addition, remote sensing data can be used to monitor environmental impacts of oil and gas production, such as land subsidence, water pollution, and deforestation.

Remote sensing is a valuable tool in the oil exploration industry, providing crucial information for locating and evaluating potential oil and gas deposits and for monitoring production and environmental impacts.

Important Questions

1. How is remote sensing utilized in oil exploration?
2. What are the different types of remote sensing techniques used in oil exploration?
3. How does satellite imagery contribute to the identification of potential oil reserves?
4. What role does LiDAR play in oil exploration?
5. How can thermal infrared imagery aid in locating oil and gas deposits?
6. What are the benefits of using hyperspectral imaging in the context of oil exploration?
7. How does gravity and magnetic field measurement assist in identifying subsurface structures for oil explorations?
8. What are the challenges associated with using remote sensing technologies in offshore oil exploration?
9. How does synthetic aperture radar contribute to mapping geological features for oil exploration?
10. What advancements have been made in remote sensing technology to enhance oil exploration techniques?

10

Engineering Geological Investigation

Remote sensing can be a useful tool in engineering geological investigations, particularly in the identification and characterization of geological features and hazards.

Some of the ways that remote sensing can be used in engineering geological investigations include:

Geological mapping: Remote sensing data can be used to create detailed maps of the geological features in an area, including rock formations, faults, and folds. This information can be used to identify potential hazards, such as landslides and rockfalls, and to plan engineering projects accordingly.

Landslide detection and monitoring: Remote sensing data, such as satellite imagery, can be used to detect landslides and monitor their movement over time. This information can help engineers to design and implement appropriate stabilization measures.

Ground deformation monitoring: Remote sensing techniques, such as Interferometric Synthetic Aperture Radar (InSAR), can be used to detect ground deformation caused by geological processes such as subsidence, uplift, and faulting. This information can be used to identify potential hazards and plan engineering projects accordingly.

Geohazard assessment: Remote sensing data can be used to identify and assess the potential for geohazards, such as landslides, earthquakes, and volcanic eruptions. This information can be used to design and implement appropriate mitigation measures.

Mineral exploration: Remote sensing data can be used to identify and map mineral deposits, which can be valuable information for engineering projects that require the extraction or utilization of minerals.

In summary, remote sensing can provide valuable information for engineering geological investigations, from identifying and mapping geological features and hazards to monitoring ground deformation and assessing the potential for geohazards.

Role of Remote Sensing in Alignment Studies

Remote sensing plays a critical role in alignment studies for infrastructure projects such as roads, tunnels, canals, and railways. Alignment studies involve the evaluation of various options for the location and alignment of the proposed infrastructure and selecting the optimal option based on factors such as cost, feasibility, environmental impacts, and safety. Remote sensing provides valuable data that can be used to assess these factors and make informed decisions.

Here are some of the ways remote sensing can be used in alignment studies

Topographic mapping: Remote sensing techniques, such as LiDAR, photogrammetry, and satellite imagery, can be used to create detailed topographic maps of the study area. These maps can provide accurate information on elevation, slope, and terrain features, which can be used to evaluate the feasibility of different alignment options.

Geotechnical mapping: Remote sensing data can be used to map the geology and soil properties of the study area. This information can be used to evaluate the stability and suitability of different alignment options and to design appropriate foundations and stabilization measures.

Environmental mapping: Remote sensing data can be used to map natural and human-made features such as wetlands, wildlife habitats, and cultural heritage sites. This information can be used to assess the environmental impacts of different alignment options and to design appropriate mitigation measures.

Risk assessment: Remote sensing data can be used to identify potential hazards and risks, such as landslides, floods, and earthquake zones. This information can be used to evaluate the safety of different alignment options and to design appropriate measures to minimize risks.

Construction monitoring: Remote sensing data can be used to monitor the progress and quality of construction, such as measuring the height and volume of earthworks and detecting any deviations from the design.

In summary, remote sensing is a valuable tool in alignment studies as it provides accurate and reliable data on the terrain, geology, and environment of the study area. This information can be used to evaluate the feasibility, stability, and safety of different alignment options and to design appropriate mitigation measures to minimize environmental impacts and risks.

Alignment Studies – Roads, Tunnels, Canals

Alignment studies are a crucial step in the design and construction of roads, tunnels, canals, and other infrastructure projects. These studies involve evaluating different options for the location and alignment of the proposed infrastructure, and selecting the optimal alignment based on a range of factors, such as cost, feasibility, environmental impacts, and safety.

Remote sensing can play an important role in alignment studies by providing detailed and accurate data on the terrain and surrounding environment. Some of the ways that remote sensing can be used in alignment studies include:

Topographic mapping: Remote sensing techniques, such as LiDAR and aerial photogrammetry, can be used to create detailed topographic maps of the proposed alignment. This information can be used to identify potential challenges and opportunities, such as steep slopes, water bodies, and vegetation cover.

Geotechnical mapping: Remote sensing data can be used to map the geology and soil properties of the area. This information can be used to evaluate the feasibility and stability of different alignment options, and to design appropriate foundations and stabilization measures.

Environmental mapping: Remote sensing data can be used to map the natural and human-made features in the area, such as wetlands, wildlife habitats, and cultural heritage sites. This information can be used to assess the environmental impacts of different alignment options and to design appropriate mitigation measures.

Risk assessment: Remote sensing data can be used to identify potential hazards and risks, such as landslides, floods, and earthquake zones. This information can be used to evaluate the safety of different alignment options and to design appropriate measures to minimize risks.

Construction monitoring: Remote sensing data can be used to monitor the progress and quality of construction, such as measuring the height and volume of earthworks, and detecting any deviations from the design.

In summary, remote sensing can provide valuable data for alignment studies, from topographic and geotechnical mapping to environmental mapping and risk assessment, and can help ensure the safe and efficient design and construction of roads, tunnels, canals, and other infrastructure projects.

Site selection studies – Dams, bridges, highways :

Dams

ite selection studies for dams involve evaluating potential locations for constructing a dam based on various factors such as geology, hydrology, environmental impact, and socio-economic considerations. These studies are essential to identify suitable sites that maximize the benefits of dam construction while minimizing potential risks and negative impacts.

Here are some key aspects considered in site selection studies for dams

Geology and Geotechnical Assessment: The geology of the area is evaluated to determine the suitability of the foundation for dam construction. Geological studies assess the stability of the rock or soil formations, potential for seepage or leakage, and the presence of faults or other geological hazards that may affect dam integrity.

Hydrological Assessment: The hydrological characteristics of the site, including rainfall patterns, river flow rates, and sedimentation rates, are analyzed to estimate the availability of water resources and the potential for reliable water storage. This information helps determine the dam's capacity and design.

Environmental Impact Assessment: Site selection studies consider the potential environmental impacts of dam construction on the surrounding ecosystem. This involves evaluating the effects on fish and wildlife habitats, vegetation, water quality, and downstream water flow regimes. Mitigation measures may be identified to minimize adverse impacts and preserve ecological balance.

Socio-Economic Considerations: The social and economic aspects of dam construction are evaluated, taking into account factors such as population displacement, resettlement, land acquisition, and potential impacts on local communities. Socio-economic studies also assess the benefits of the dam in terms of water supply, flood control, hydropower generation, irrigation, and navigation.

Dam Safety and Risk Assessment: Site selection studies involve assessing potential risks associated with dam construction and operation. This includes evaluating seismic hazards, flood risks, and the stability of reservoir slopes. Mitigation measures are identified to ensure the safety and stability of the dam throughout its lifespan.

Cost and Feasibility Analysis: Economic feasibility studies analyze the financial viability of dam construction, considering the estimated costs of design, construction, operation, and maintenance. Cost-benefit analysis is performed to evaluate the economic benefits derived from the dam against the investment required.

Stakeholder Engagement: Site selection studies involve engaging with various stakeholders, including local communities, government agencies, environmental organizations, and indigenous populations. Public consultations and participation help gather input, address concerns, and incorporate local perspectives into the decision-making process.

Site selection studies for dams are comprehensive assessments that consider technical, environmental, social, and economic factors to identify suitable locations for dam construction while minimizing adverse impacts and maximizing the benefits for both human and natural systems.

Bridges

Site selection studies for bridges involve evaluating potential locations for constructing a bridge based on various factors such as geography, transportation needs, environmental considerations, and structural feasibility. These studies are essential to identify suitable sites that ensure efficient and safe connectivity while minimizing environmental impact.

Here are some key aspects considered in site selection studies for bridges

Transportation Needs and Traffic Analysis: The study begins by assessing the transportation needs of the area, including current and projected traffic volumes, connectivity requirements, and the type of vehicles that will use the bridge. Traffic analysis helps determine the appropriate location for the bridge to facilitate efficient movement of goods, people, and vehicles.

Geographic Considerations: The geography of the area plays a crucial role in bridge site selection. Factors such as water bodies, topography, geological conditions, and existing infrastructure are evaluated. The presence of rivers, lakes, or other water bodies may necessitate the need for a bridge, and the geological stability of the site is assessed to ensure a solid foundation for the bridge.

Environmental Impact Assessment: Environmental considerations are vital in bridge site selection studies. The impact on natural habitats, protected areas, wetlands, and sensitive ecosystems is evaluated. Mitigation measures are identified to minimize disruption and preserve biodiversity. The potential impact on water quality, sedimentation, and wildlife corridors is also considered.

Structural Feasibility and Engineering Considerations: The structural feasibility of a bridge is assessed based on factors such as span length, foundation conditions, and the need for any special features like high piers or cable-stayed structures. Structural engineers analyze the site to ensure that the proposed bridge design can accommodate the anticipated loads, including traffic, wind, and seismic forces.

Cost and Construction Feasibility: The cost of bridge construction and associated infrastructure is evaluated to assess the project's feasibility. Factors such as land acquisition, construction techniques, material availability, and potential construction challenges are considered. Cost estimates are prepared to evaluate the economic viability of the proposed bridge.

Socio-economic Considerations: The social and economic impacts of the bridge on the surrounding communities and economy are analyzed. This includes evaluating factors such as access to markets, employment opportunities, tourism potential, and the impact on local communities. The bridge should benefit the region by improving transportation efficiency and promoting economic development.

Stakeholder Engagement: Site selection studies for bridges involve engaging with various stakeholders, including local communities, transportation authorities, environmental organizations, and government agencies. Public consultations are conducted to gather input, address concerns, and incorporate local perspectives into the decision-making process.

By considering these aspects, site selection studies for bridges aim to identify optimal locations that provide safe and efficient transportation solutions while minimizing environmental impact and promoting sustainable development.

Highways

Site selection studies for highways involve evaluating potential routes for constructing new highways or expanding existing ones. These studies assess various factors such as transportation needs, environmental impact, land use, engineering feasibility, and socio-economic considerations. The goal is to identify the most suitable alignment for the highway that balances the needs of efficient transportation with minimal environmental and social disruption.

Here are key aspects considered in site selection studies for highways

Transportation Needs and Traffic Analysis: The study begins by evaluating the existing transportation network and identifying the need for new or improved highways. Traffic analysis is conducted to determine current and projected traffic volumes, travel patterns, and congestion points. This information helps identify potential routes that address transportation needs and improve connectivity.

Environmental Impact Assessment: Environmental considerations play a crucial role in highway site selection. The study assesses potential impacts on natural habitats, protected areas, wetlands, and sensitive ecosystems. It also evaluates the impact on air and water quality, noise levels, and visual aesthetics. Mitigation measures are identified to minimize environmental disruption and preserve ecological balance.

Land Use and Land Acquisition: The study evaluates the existing land use patterns and assesses the availability of suitable land for the proposed highway alignment. It considers factors such as the impact on agricultural land, residential areas, cultural or historical sites, and potential conflicts with existing infrastructure. Land acquisition requirements and associated costs are also considered.

Engineering Feasibility and Geotechnical Assessment: The engineering feasibility of potential highway alignments is evaluated. This involves analyzing topographic conditions, geological features, soil stability, and hydrological considerations. Geotechnical studies assess the suitability of the ground for highway construction and identify any potential geotechnical hazards.

Socio-economic Considerations: The study assesses the social and economic impacts of the proposed highway. It evaluates factors such as accessibility to employment centers, impact on local communities and businesses, potential for economic development, and connectivity to key facilities like schools, hospitals, and markets. Social and economic benefits and potential adverse effects are taken into account.

Cost Analysis and Financial Feasibility: Cost estimates are prepared to assess the financial viability of the proposed highway. This includes evaluating construction costs, land acquisition expenses, operation and maintenance costs, and potential revenue streams. Cost-benefit analysis is conducted to weigh the economic benefits of the highway against the investment required.

Stakeholder Engagement: Site selection studies for highways involve engaging with various stakeholders, including local communities, transportation authorities, environmental organizations, and government agencies. Public consultations and stakeholder meetings are conducted to gather input, address concerns, and incorporate local perspectives into the decision-making process.

By considering these factors, site selection studies for highways aim to identify optimal routes that improve transportation efficiency, promote economic development, minimize environmental impact, and address the needs of the local communities and stakeholders.

Important Questions

1. What is the geological composition of the project site and how might it impact construction?
2. What are the potential geological hazards (e.g., landslides, sinkholes) in the area and how can they be mitigated?

3. What is the depth and quality of the bedrock at the project site, and how might it affect foundation design?
4. Are there any known fault lines or seismic activity in the vicinity of the project site tat could impact the structural integrity of the proposed construction?
5. What is the groundwater table level at the site, and how might it influence excavation and dewatering processes?
6. Are there any underground utilities or geological features (e.g., caves, voids) that could pose challenges during construction?
7. What is the soil classification and bearing capacity at the site, and how will it influence the design and stability of structures?
8. How has the site's geological history affected the formation and distribution of soil and rock layers present?
9. Are there any specific environmental considerations related to the project site, such as protected ecosystems or sensitive habitats, that need to be accounted for during construction?
10. What is the anticipated long-term behaviour of the geological formations at the project site, and how might it impact the maintenance and lifespan of the proposed structures?

11

Natural Disaster Mapping and Management

Definition of Disaster

A disaster can be defined as a sudden or catastrophic event that causes significant damage, destruction, and disruption to human life, property, or the environment. Disasters can be natural, such as earthquakes, hurricanes, floods, wildfires, or they can be human-made, including industrial accidents, terrorist attacks, or technological failures.

Disasters often result in the loss of lives, injuries, displacement of people, infrastructure damage, and the interruption of essential services like electricity, water, and communication systems. They can have long-lasting impacts on communities, economies, and the environment.

Governments, organizations, and communities implement disaster preparedness, response, and recovery strategies to mitigate the effects of disasters, protect lives and property, and facilitate the rebuilding and restoration of affected areas.

Types of Disaster

There are various types of disasters that can occur. Here are some of the most common types:

1. Natural Disasters

- Earthquakes
- Hurricanes, typhoons, and cyclones
- Floods
- Tornadoes
- Droughts
- Wildfires
- Landslides
- Avalanches

- Tsunamis
- Volcanic eruptions

2. Weather-Related Disasters

- Severe storms
- Blizzards
- Hailstorms
- Heatwaves
- Cold waves

3. Human-Made Disasters

- Industrial accidents
- Chemical spills
- Nuclear accidents
- Terrorist attacks
- Infrastructure failures (e.g., bridge collapses)
- Civil unrest and conflicts
- Transportation accidents (e.g., plane crashes)

4. Health-Related Disasters

- Pandemics and epidemics
- Disease outbreaks
- Biological or chemical warfare

5. Environmental Disasters

- Oil spills
- Contamination of water sources
- Deforestation
- Soil erosion
- Desertification

These are just a few examples, and there are other types of disasters that can occur. Each type of disaster has its unique characteristics, causes, and impacts, requiring specific response and recovery strategies.

Disaster in India

India, being a geographically diverse country, is prone to various types of disasters. Here are some of the significant disasters that have occurred in India:

1. Cyclones

- Cyclone Amphan (2020)
- Cyclone Fani (2019)
- Cyclone Hudhud (2014)
- Cyclone Phailin (2013)
- Cyclone Gaja (2018)
- Cyclone Nisarga(2020)

2. Floods

- Kerala floods (2018)
- Uttarakhand floods (2013)
- Chennai floods (2015)
- Bihar floods (2020)
- Assam floods (annual occurrence)

3. Earthquakes

- Gujarat earthquake (2001)
- Bhuj earthquake (2001)
- Sikkim earthquake (2011)
- Nepal earthquake (2015)

4. Heatwaves

- Heatwave in Andhra Pradesh and Telangana (2015)
- Heatwave in North India (2019)
- Heatwave in Maharashtra (2021)

5. Landslides

- Malin landslide, Maharashtra (2014)
- Kedarnath landslides, Uttarakhand (2013)
- Munnar landslides, Kerala (2020)

6. Industrial Accidents

- Bhopal gas tragedy (1984)
- Visakhapatnam gas leak (2020)
- NTPC Unchahar plant explosion (2017)

7. Droughts

- Maharashtra drought (2019)
- Tamil Nadu drought (ongoing)
- Bundelkhand drought (ongoing)

8. Terrorist Attacks

- Mumbai terrorist attacks (2008)
- Pathankot airbase attack (2016)
- Pulwama attack (2019)

9. Epidemics and Pandemics

- COVID-19 pandemic (ongoing)
- Swine flu outbreak (2009)

10. Forest Fires

- Uttarakhand forest fires (2016)
- Himachal Pradesh forest fires (2019)

These are just a few examples, and India has faced numerous other disasters over the years. The Indian government, along with various agencies and organizations, works towards disaster preparedness, response, and recovery to mitigate the impact of such disasters and provide relief and support to affected communities.

Important Case Study of Disaster

Case Study 1

Cyclone Amphan was a powerful tropical cyclone that struck parts of India and Bangladesh in May 2020. It was one of the strongest cyclones to have ever formed in the Bay of Bengal. Here are some key details and impacts of Cyclone Amphan:

Formation and Intensity: Cyclone Amphan originated from a low-pressure area over the Southeast Bay of Bengal on May 16, 2020. It rapidly intensified into a super cyclonic storm, reaching a peak intensity with maximum sustained winds of about 185 km/h (115 mph) and gusts up to 240 km/h (150 mph).

Affected Regions: Cyclone Amphan made landfall on May 20, 2020, near the border of India and Bangladesh. The most severely affected regions in India included West Bengal and Odisha, with Kolkata, the capital of West Bengal, being significantly impacted.

Devastating Impact: Cyclone Amphan caused widespread destruction and had a devastating impact on the affected regions. It led to extensive damage to infrastructure, including houses, buildings, roads, and power lines. Large-scale uprooting of trees and flooding occurred, exacerbating the destruction.

Loss of Lives and Displacement: The cyclone resulted in the loss of several lives, with casualties reported in both India and Bangladesh. Many people were displaced from their homes and sought shelter in evacuation centers and relief camps.

Agricultural and Economic Losses: Cyclone Amphan caused significant damage to agriculture, particularly in the Sundarbans region, affecting crops, livestock, and fishing activities. The cyclone also impacted the economy of the affected areas, including businesses, industries, and transportation networks.

Relief and Rehabilitation Efforts: The Indian and Bangladeshi governments, along with humanitarian organizations, conducted extensive relief and rehabilitation efforts to provide immediate assistance to affected communities. These efforts included providing food, clean water, medical aid, and restoring essential services.

Climate Change and Preparedness: Cyclone Amphan highlighted the vulnerability of coastal regions to the impacts of climate change. It emphasized the need for improved disaster preparedness, early warning systems, and resilient infrastructure to mitigate the effects of future cyclones.

Cyclone Amphan serves as a reminder of the devastating power of tropical cyclones and the importance of disaster preparedness and response in mitigating their impact on vulnerable regions.

Case Study 2

Cyclone Nisarga was a severe cyclonic storm that made landfall in Maharashtra, India, in June 2020. It was one of the rare cyclones to affect the state's coastline. Here are some key details and impacts of Cyclone Nisarga:

Formation and Intensity: Cyclone Nisarga originated from a low-pressure area over the Arabian Sea on June 1, 2020. It gradually intensified into a severe cyclonic storm with maximum sustained winds of about 100-110 km/h (60-70 mph) and gusts up to 120 km/h (75 mph).

Affected Regions: Cyclone Nisarga primarily impacted the coastal regions of Maharashtra, including Mumbai, the capital city. Other areas affected included Raigad, Palghar, Thane, and Ratnagiri districts.

Impact on Mumbai: Mumbai, a densely populated city, experienced strong winds and heavy rainfall as the cyclone approached. The city was put on high alert, and precautionary measures were taken to ensure the safety of the population.

Damage to Infrastructure: Cyclone Nisarga caused damage to infrastructure, including houses, buildings, power lines, and trees. The strong winds led to the uprooting of trees and disrupted electricity supply in some areas.

Evacuation and Rescue Efforts: The authorities carried out extensive evacuation and rescue operations to move people from vulnerable areas to safer locations. Evacuation centers and relief camps were set up to provide temporary shelter and necessary support.

Minimal Loss of Life: Thanks to the preparedness measures and timely evacuation efforts, the loss of life was relatively low. However, a few casualties were reported, and there were instances of injuries caused by falling trees and debris.

Aftermath and Rehabilitation: Following the cyclone, restoration work was initiated to repair damaged infrastructure and restore essential services. Efforts were made to provide relief materials, food, and other necessary assistance to affected communities.

Cyclone Nisarga served as a reminder of the vulnerability of coastal regions to cyclonic storms and emphasized the importance of disaster preparedness and early warning systems in saving lives and minimizing damage. It highlighted the need for ongoing efforts to enhance resilience and response capabilities in coastal areas prone to cyclones.

Case Study 3

The Kerala floods of 2018 were a devastating natural disaster that occurred in the southern Indian state of Kerala. The floods, which were triggered by unusually heavy rainfall during the monsoon season, resulted in widespread flooding, landslides, and destruction across the state. Here are some key details about the Kerala floods:

Timing: The floods occurred between late May and early August 2018, with the worst period of flooding happening in late July and early August.

Rainfall: Kerala experienced extremely heavy rainfall during this period, significantly exceeding the average rainfall for the monsoon season. Several districts received rainfall that was more than double the usual amount, leading to severe flooding.

Impact: The floods affected almost the entire state of Kerala, with over 1.4 million people displaced from their homes and seeking refuge in relief camps. The flooding and landslides caused extensive damage to infrastructure, including roads, bridges, buildings, and agricultural fields.

Casualties: The floods resulted in a tragic loss of life. Official reports estimate that over 400 people lost their lives during the disaster. Many others were injured or reported missing.

Rescue and Relief Operations: The Indian government, along with various state agencies, the armed forces, and non-governmental organizations (NGOs), launched massive rescue and relief operations to evacuate stranded individuals and provide aid to the affected population. Helicopters, boats, and other resources were deployed for rescue operations.

Economic Impact: The floods had a significant economic impact on Kerala. The agricultural sector suffered extensive damage, leading to loss of crops and livestock. Businesses and tourism were also severely affected, as many areas were inaccessible or damaged.

Unity and Support: The Kerala floods saw an overwhelming response from the public, both within the state and from across the country. People came together to provide relief materials, volunteer services, and financial aid. Social media played a crucial role in mobilizing support and coordinating relief efforts.

Post-Flood Rehabilitation: After the floodwaters receded, the focus shifted to the rehabilitation and rebuilding phase. The government implemented various measures to provide assistance and support to the affected individuals and communities, including financial aid, housing reconstruction, and infrastructure repair.

The Kerala floods of 2018 were one of the worst natural disasters in the history of the state. The resilience and solidarity displayed by the people of Kerala, along with the support from various organizations and the government, played a significant role in the recovery and rebuilding process.

Case Study 4

The Bihar floods of 2020 were another devastating natural disaster that occurred in the eastern Indian state of Bihar. These floods were a result of heavy rainfall during the monsoon season and affected several districts in the state. Here's an overview of the Bihar floods in 2020:

Timing: The floods occurred between June and October 2020, with the peak period of flooding in July and August.

Rainfall and River Overflow: Bihar experienced excessive rainfall during this period, causing major rivers like the Ganga, Sone, and Kosi to overflow their banks. This led to widespread flooding across the affected districts.

Impact: The floods affected millions of people in Bihar. Thousands of villages were submerged, and extensive damage was caused to infrastructure, including roads, bridges, and houses. The flooding also led to the loss of crops, livestock, and disrupted normal life.

Casualties: The floods resulted in a significant loss of life. According to official reports, over 130 people lost their lives during the floods, and many others were injured or displaced.

Relief and Rescue Operations: The state government, along with central agencies and NGOs, initiated rescue and relief operations to evacuate affected individuals and provide assistance. The National Disaster Response Force (NDRF) and State Disaster Response Force (SDRF) were deployed for rescue operations, while relief camps were set up to provide shelter and essential supplies.

Health Concerns: Floods often lead to health-related challenges. Waterborne diseases, such as cholera and typhoid, pose a threat due to contamination of water sources. The state government took measures to provide medical assistance and ensure clean drinking water to prevent outbreaks.

Rehabilitation and Recovery: After the floodwaters receded, the focus shifted to rehabilitation and recovery efforts. The government provided financial aid, distributed relief materials, and undertook infrastructure repair and reconstruction projects to help affected individuals and communities.

Long-term Measures: The Bihar government, along with relevant authorities, has been working on long-term measures to mitigate the impact of floods. This includes measures such as constructing embankments, improving drainage systems, and implementing early warning systems to minimize damage from future flood events.

The Bihar floods of 2020 highlighted the vulnerability of the state to flooding during the monsoon season. Efforts are being made to improve preparedness and response mechanisms to mitigate the impact of future flood events and ensure the safety and well-being of the affected population.

Case Study 5

The Bhuj earthquake, also known as the Gujarat earthquake, occurred on January 26, 2001, in the state of Gujarat, India. It was one of the most devastating earthquakes to hit India in recent history. Here's an overview of the Bhuj earthquake:

Magnitude and Epicenter: The earthquake had a magnitude of 7.7 on the Richter scale. Its epicenter was near the town of Bhuj in Kutch district, Gujarat.

Timing and Impact: The earthquake struck in the early morning hours, around 8:46 am local time. It caused widespread destruction across a large area, affecting several districts in Gujarat, including Bhuj, Anjar, Bhachau, and Rapar.

Casualties and Damage: The earthquake resulted in a significant loss of life and extensive damage. Official reports estimate that over 20,000 people lost their lives, and more than 167,000 were injured. Numerous buildings, including residential structures, schools, hospitals, and infrastructure such as roads and bridges, were severely damaged or collapsed.

Rescue and Relief Operations: Following the earthquake, rescue and relief operations were launched on a massive scale. The Indian government, along with international aid agencies and the armed forces, deployed personnel, equipment, and resources for search and rescue efforts, medical assistance, and providing relief supplies to the affected population.

Rehabilitation and Reconstruction: After the immediate rescue and relief efforts, the focus shifted to long-term rehabilitation and reconstruction. The government initiated various measures to provide housing, financial assistance, and infrastructure redevelopment for the affected communities. The rebuilding process was extensive and took several years to complete.

Economic Impact: The Bhuj earthquake had a significant economic impact on the region. Industries, including manufacturing and agriculture, were disrupted, leading to a decline in production and livelihoods. The government implemented measures to support economic recovery and development in the affected areas.

Lessons Learned and Preparedness: The earthquake highlighted the need for improved disaster preparedness and response mechanisms in India. It led to the implementation of stricter building codes and guidelines for earthquake-resistant structures. Efforts were made to increase public awareness about earthquake safety and strengthen disaster management systems.

The Bhuj earthquake of 2001 remains a tragic event in the history of India, with its profound impact on human lives, infrastructure, and the economy. The disaster prompted significant efforts to rebuild and develop more resilient communities, ensuring that lessons learned from the event contribute to better preparedness for future earthquakes.

Case Study 6

The Nepal earthquake, also known as the Gorkha earthquake, occurred on April 25, 2015, in Nepal, a landlocked country in South Asia. It was one of the deadliest earthquakes in the region in recent history. Here's an overview of the Nepal earthquake:

Magnitude and Epicenter: The earthquake had a magnitude of 7.8 on the Richter scale. Its epicenter was near the village of Barpak in the Gorkha district, about 80 kilometers northwest of the capital city, Kathmandu.

Timing and Impact: The earthquake struck in the mid-morning hours, around 11:56 am local time. It caused widespread destruction across Nepal, affecting numerous districts, including Kathmandu Valley and regions in the western and central parts of the country.

Casualties and Damage: The earthquake resulted in a significant loss of life and extensive damage. Official reports estimate that over 9,000 people lost their lives, and more than 22,000 were injured. The earthquake caused the collapse of numerous buildings, including historical monuments, temples, and residential structures. Infrastructure such as roads, bridges, and hospitals were also severely damaged.

Aftershocks: Following the main earthquake, Nepal experienced numerous aftershocks, some of which were also of significant magnitude. These aftershocks caused further damage and hampered rescue and relief efforts.

Rescue and Relief Operations: After the earthquake, immediate rescue and relief operations were launched. The Nepalese government, along with international aid agencies, neighboring countries, and the armed forces, mobilized personnel, equipment, and resources for search and rescue efforts, medical assistance, and providing relief supplies to the affected population. The difficult terrain and remote locations posed challenges for access and distribution of aid.

Rehabilitation and Reconstruction: The focus shifted to long-term rehabilitation and reconstruction after the immediate rescue and relief efforts. The Nepalese government, along with international support, initiated measures to rebuild homes, infrastructure, and cultural heritage sites. The process of recovery and reconstruction has been ongoing and continues to this day.

Economic Impact: The Nepal earthquake had a significant economic impact on the country. Tourism, a vital industry for Nepal, was severely affected due to the damage to cultural heritage sites and infrastructure. The disruption to agriculture, trade, and businesses further contributed to economic challenges.

Preparedness and Risk Mitigation: The earthquake served as a wake-up call for Nepal and highlighted the need for improved disaster preparedness and risk mitigation strategies. Efforts have been made to strengthen building codes, enhance early warning systems, and increase public awareness about earthquake safety.

The Nepal earthquake of 2015 was a devastating event that caused immense loss of life and widespread destruction. The recovery and rebuilding process in Nepal has been a long-term endeavor, with efforts focused on reconstruction, infrastructure development, and disaster risk reduction to build a more resilient nation.

Case Study 7

In 2019, North India experienced an intense heatwave during the summer months, particularly in June. The heatwave resulted in scorching temperatures and adverse effects on human health, agriculture, and daily life. Here's an overview of the heatwave in North India in 2019:

Duration and Intensity: The heatwave persisted for several weeks, with the most extreme conditions occurring in June. Temperatures soared well above normal, reaching record-breaking levels in some areas. Many places recorded temperatures exceeding 45 degrees Celsius (113 degrees Fahrenheit).

Affected Regions: The heatwave primarily affected the northern states of India, including Rajasthan, Uttar Pradesh, Haryana, Delhi, Punjab, and parts of Madhya Pradesh. These regions experienced exceptionally high temperatures and prolonged periods of hot weather.

Health Impacts: The heatwave had significant health consequences. Heat-related illnesses, such as heatstroke, dehydration, heat exhaustion, and respiratory problems, were widespread. Hospitals and medical facilities faced an influx of patients seeking treatment for heat-related ailments.

Agricultural Impact: The heatwave had a severe impact on agriculture. Crops, particularly those sensitive to extreme heat, were damaged or failed to grow. Irrigation systems struggled to cope with the increased demand for water, exacerbating the agricultural challenges.

Power Outages: The soaring temperatures resulted in increased electricity consumption, leading to power outages in some areas. Frequent power cuts added to the discomfort and challenges faced by the population during the heatwave.

Public Health Measures: Authorities implemented various measures to mitigate the effects of the heatwave. Public advisories were issued, urging people to stay hydrated, avoid unnecessary exposure to the sun, and take precautions to prevent heat-related illnesses. Cooling centers and shelters were set up to provide respite to those without access to air conditioning.

Climate Change Considerations: The heatwave in North India in 2019 was consistent with the global trend of increasing heatwaves, which is attributed to climate change. Rising temperatures and changes in weather patterns contribute to the frequency and intensity of such extreme heat events.

Preparedness and Resilience: The heatwave highlighted the need for improved preparedness and resilience to cope with extreme weather events. It emphasized the importance of urban planning, infrastructure development, and public health strategies to mitigate the impact of heatwaves on vulnerable populations.

The 2019 heatwave in North India served as a stark reminder of the risks posed by extreme heat and the urgent need to address climate change and adapt to its consequences. It underscored the importance of measures to protect public health, enhance infrastructure, and promote sustainable practices to build resilience in the face of rising temperatures.

Case Study 8

In 2021, the state of Maharashtra in India experienced a severe heatwave during the summer months. The heatwave brought scorching temperatures and had significant impacts on the region. Here's an overview of the heatwave in Maharashtra in 2021:

Duration and Intensity: The heatwave persisted for an extended period, with the most intense conditions occurring in the months of April, May, and June. Temperatures soared well above normal, reaching extreme levels in many parts of the state.

Affected Regions: The heatwave affected various parts of Maharashtra, including major cities like Mumbai, Pune, Nagpur, and Aurangabad. These regions experienced prolonged periods of extremely hot weather.

High Temperatures: Maharashtra witnessed temperatures exceeding 45 degrees Celsius (113 degrees Fahrenheit) in several places. Some areas even recorded temperatures surpassing 47 degrees Celsius (116.6 degrees Fahrenheit), making it one of the hottest summers in the region's recent history.

Health Impacts: The heatwave had significant health consequences for the population. Heat-related illnesses, including heat exhaustion and heatstroke, were widespread. Hospitals and medical facilities faced an increased burden as people sought treatment for heat-related ailments.

Water Shortages: The extreme heat and prolonged dry spells exacerbated water scarcity in Maharashtra. The soaring temperatures increased water demand for drinking, sanitation, and agriculture, leading to water shortages in many areas.

Power Supply Challenges: The excessive electricity consumption during the heatwave caused strain on the power supply. Some regions experienced power outages due to the increased demand for air conditioning and cooling systems.

Agricultural Impact: The heatwave had adverse effects on agriculture. Crops and livestock suffered due to water scarcity and extreme heat. Crop yields were affected, and farmers faced significant challenges in managing their agricultural activities.

Mitigation Measures: Authorities took various measures to mitigate the impact of the heatwave. Public advisories were issued to raise awareness about heat-related health risks, and precautions were recommended. Efforts were made to provide relief and distribute drinking water to affected areas.

Climate Change Considerations: The heatwave in Maharashtra in 2021 aligns with the global trend of increasing heatwaves, which is attributed to climate change. Rising temperatures and changing weather patterns contribute to the frequency and intensity of extreme heat events.

Preparedness and Resilience: The heatwave emphasized the need for improved preparedness and resilience to cope with extreme weather events. It highlighted the importance of implementing sustainable practices, enhancing urban planning, and investing in climate adaptation strategies to reduce vulnerability to heatwaves.

The heatwave in Maharashtra in 2021 underscored the risks associated with extreme heat and the urgency to address climate change. It emphasized the need for comprehensive measures to protect public health, enhance water management systems, and build resilience to heatwaves and other climate-related challenges.

Case Study 9

The Malin landslide refers to a tragic natural disaster that occurred on July 30, 2014, in the village of Malin in the Ambegaon taluka of Pune district, Maharashtra, India. It resulted in a massive landslide that engulfed the entire village, causing widespread devastation and loss of life. Here are some key details about the Malin landslide:

Triggering Event: The landslide was triggered by heavy rainfall in the region, which saturated the soil and destabilized the hillsides. The hilly terrain, combined with the intensity of the rainfall, led to the catastrophic landslide.

Impact: The landslide engulfed the entire village of Malin, burying houses and structures under a massive amount of debris. The event resulted in significant loss of life and caused extensive damage to property and infrastructure.

Casualties: The Malin landslide resulted in the tragic loss of lives. Official reports indicate that more than 150 people, including men, women, and children, were buried under the debris. Rescue efforts were initiated to locate and recover survivors, but the scale of the disaster made it challenging.

Rescue and Relief Operations: Following the landslide, rescue teams from the National Disaster Response Force (NDRF), local authorities, and volunteers were mobilized for search and rescue operations. They worked tirelessly to locate survivors and recover bodies. Relief efforts were also undertaken to provide aid, shelter, and support to the affected families.

Rehabilitation and Recovery: After the rescue operations, the focus shifted to the rehabilitation and recovery of the affected community. Efforts were made to provide housing, financial assistance, and psychological support to the survivors. The government and various organizations were involved in rebuilding and restoring the village.

Lessons Learned: The Malin landslide served as a reminder of the vulnerability of hilly areas to landslides, especially during periods of heavy rainfall. It highlighted the importance of early warning systems, proper land-use planning, and adequate infrastructure measures to minimize the risks associated with landslides.

The Malin landslide in Maharashtra in 2014 was a devastating event that resulted in the loss of many lives and had a profound impact on the affected community. Efforts were made to provide support and aid to the survivors and implement measures to prevent such disasters in the future.

Case Study 10

The Kedarnath landslides of 2013 refers to a catastrophic natural disaster that occurred in the state of Uttarakhand, India. The event involved heavy rainfall, flash floods, and landslides that severely impacted the Kedarnath region, including the famous Kedarnath Temple and surrounding areas. Here are some key details about the Kedarnath landslides:

Triggering Event: The disaster was triggered by heavy rainfall in the region during the monsoon season. The excessive rainfall resulted in flash floods and landslides in the mountainous terrain of Uttarakhand.

Impact: The landslides and flash floods had a devastating impact on the Kedarnath region. The flooding caused extensive damage to infrastructure, including roads, buildings, and bridges. The Kedarnath Temple complex and nearby areas were significantly affected.

Casualties: The disaster resulted in a tragic loss of life. The exact number of casualties is uncertain, but it is estimated that thousands of people, including pilgrims and residents, lost their lives in the floods and landslides.

Rescue and Relief Operations: Following the disaster, massive rescue and relief operations were launched. The Indian government, along with the armed forces, disaster response teams, and local volunteers, initiated efforts to rescue stranded individuals, provide medical aid, and supply food and essential supplies to those affected.

Environmental Impact: The landslides and floods had a significant environmental impact. The loss of vegetation and soil erosion in the affected areas disrupted the fragile mountain ecosystem. It also led to concerns of long-term impacts on the region's biodiversity and water resources.

Rehabilitation and Recovery: After the immediate rescue and relief efforts, the focus shifted to the rehabilitation and recovery phase. Efforts were made to rebuild infrastructure, restore services, and provide support to the affected communities. The restoration of the Kedarnath Temple complex was also undertaken.

Lessons Learned: The Kedarnath landslides served as a wake-up call regarding the vulnerability of the region to extreme weather events and the importance of disaster preparedness and mitigation measures. The disaster highlighted the need for improved land-use planning, early warning systems, and infrastructure resilience in mountainous areas prone to landslides and flash floods.

The Kedarnath landslides and floods of 2013 were a catastrophic event that caused immense loss of life and infrastructure damage. The recovery and rebuilding process has been ongoing, focusing on restoring the affected region and implementing measures to reduce the vulnerability of Uttarakhand to similar disasters in the future.

Case Study 11

The Visakhapatnam gas leak refers to a tragic incident that occurred on May 7, 2020, in Visakhapatnam, Andhra Pradesh, India. A gas leak took place at the LG Polymers chemical plant, resulting in the release of toxic styrene gas into the surrounding area. Here are some key details about the Visakhapatnam gas leak:

Incident: The gas leak occurred in the early morning hours at the LG Polymers chemical plant located in the RR Venkatapuram village near Visakhapatnam. The leak was caused by a styrene gas storage tank.

Toxic Gas: Styrene gas, a highly flammable and toxic substance, leaked from the storage tank. Styrene is used in the production of polystyrene and other plastic products.

Impact: The gas leak resulted in the release of a large quantity of toxic gas into the surrounding atmosphere. The gas spread quickly, affecting nearby residential areas and villages.

Casualties and Injuries: The gas leak led to the tragic loss of several lives. Official reports indicate that at least 12 people lost their lives, and hundreds of others were hospitalized due to exposure to the toxic gas. Many experienced symptoms such as respiratory distress, eye irritation, and nausea.

Evacuation and Rescue Operations: As news of the gas leak spread, authorities initiated evacuation measures to move residents out of the affected areas. Emergency response teams, including the National Disaster Response Force (NDRF) and local authorities, were deployed for rescue operations and to provide medical assistance to the affected individuals.

Investigations and Legal Proceedings: Following the incident, investigations were conducted to determine the cause of the gas leak and assess the responsibility of the chemical plant. Legal proceedings were initiated to hold those accountable for the incident.

Environmental Impact: The gas leak also had environmental consequences. The toxic gas release affected air quality in the region and posed risks to ecosystems and wildlife.

Compensation and Rehabilitation: The affected individuals and families were provided with compensation for the loss of lives, injuries, and damages. Rehabilitation measures were undertaken to support the affected communities and help them recover from the incident.

The Visakhapatnam gas leak in 2020 was a tragic incident that resulted in loss of lives, injuries, and environmental damage. The incident highlighted the importance of ensuring the safety of industrial facilities and the need for stringent regulations and protocols to prevent such accidents in the future.

Case Study 12

The NTPC Unchahar plant explosion refers to a major industrial accident that occurred on November 1, 2017, at the NTPC (National Thermal Power Corporation) thermal power plant in Unchahar, Uttar Pradesh, India. The explosion took place in a boiler unit, resulting in a significant loss of life and causing extensive damage to the plant. Here are some key details about the NTPC Unchahar plant explosion:

Incident: The explosion occurred in the coal-fired boiler unit of the NTPC Unchahar thermal power plant. The exact cause of the explosion was determined to be a rupture in a high-pressure steam pipe.

Impact: The explosion resulted in a massive release of steam, hot ash, and gases, leading to a catastrophic event at the plant. The explosion caused significant damage to the boiler unit and surrounding infrastructure.

Casualties: The incident resulted in a tragic loss of life. Official reports state that at least 45 workers were killed, and many others sustained severe injuries. The victims were primarily plant workers and contract laborers.

Rescue and Relief Operations: Following the explosion, rescue and relief operations were launched to provide immediate medical assistance to the injured and evacuate the affected workers. Local authorities, along with NDRF (National Disaster Response Force) teams and other emergency response agencies, were deployed to manage the situation.

Investigations and Accountability: An investigation was initiated to determine the cause of the explosion and assess any potential lapses in safety protocols. The incident highlighted the importance of ensuring proper maintenance, adherence to safety regulations, and regular inspections at industrial plants.

Impact on Power Generation: The explosion had an impact on the power generation capacity of the NTPC Unchahar plant. The damaged boiler unit required repairs, and the plant's operations were temporarily affected.

Compensation and Rehabilitation: The families of the deceased workers were provided with compensation, and efforts were made to support the injured and their families. Rehabilitation measures were undertaken to assist the affected individuals and provide support during their recovery process.

The NTPC Unchahar plant explosion in 2017 was a tragic incident that resulted in the loss of numerous lives and had a significant impact on the power plant's operations. The incident emphasized the importance of stringent safety measures, regular maintenance, and continuous monitoring of industrial facilities to prevent such accidents and ensure the well-being of workers.

Case Study 13

The Maharashtra drought in 2019 refers to a severe drought that affected the state of Maharashtra, India. The drought was characterized by prolonged periods of low rainfall, resulting in water scarcity, agricultural losses, and adverse impacts on livelihoods. Here are some key details about the Maharashtra drought in 2019:

Rainfall Deficit: The state of Maharashtra experienced a significant deficit in rainfall during the monsoon season in 2019. Several districts in the state received below-average rainfall, leading to water scarcity and drought-like conditions.

Agricultural Impact: The drought had a severe impact on agriculture, which is a vital sector in Maharashtra. Crop failure, loss of livestock, and reduced agricultural productivity were observed. Farmers faced challenges such as lack of water for irrigation and inadequate fodder for livestock.

Water Scarcity: The drought resulted in acute water scarcity in many regions of Maharashtra. Water sources, such as rivers, lakes, and groundwater reservoirs, dried up or reached critically low levels. People faced difficulties in accessing clean drinking water for their daily needs.

Crop Failure and Economic Consequences: The drought-related crop failure had significant economic consequences. Farmers incurred heavy financial losses due to the lack of yield and decreased farm income. The agricultural distress also had ripple effects on the rural economy and livelihoods of farmers and agricultural laborers.

Migration and Social Impact: The drought forced many people, especially farmers and their families, to migrate in search of employment and better living conditions. This led to social and economic disruptions, as well as increased pressure on urban areas.

Government Interventions: The government took several measures to address the impact of the drought. These included providing financial assistance to affected farmers, implementing water conservation and management projects, distributing fodder for livestock, and deploying tankers to supply water to drought-hit regions.

Relief and Rehabilitation Efforts: Relief camps and centers were set up to provide support to affected communities. NGOs, government agencies, and local communities collaborated to distribute food, water, and other essential supplies to those in need.

Long-Term Solutions: The drought highlighted the need for long-term solutions to mitigate the impact of water scarcity in Maharashtra. These include investment in water conservation projects, watershed management, promotion of efficient irrigation practices, and diversification of livelihood options.

The Maharashtra drought in 2019 underscored the vulnerability of the region to climate variability and the importance of sustainable water management practices. It emphasized the need for resilience-building measures and policies to reduce the impact of droughts on agriculture, water resources, and rural communities.

Case Study 14

Bundelkhand is a region in central India that spans across parts of Uttar Pradesh and Madhya Pradesh states. It has been facing a persistent and recurring drought situation for several years. The Bundelkhand drought is an ongoing issue with severe implications for agriculture, water availability, and the livelihoods of the local population. Here are some key details about the Bundelkhand drought:

Rainfall Deficit: The region of Bundelkhand has been experiencing a consistent deficit in rainfall for many years. Irregular and insufficient rainfall during the monsoon season has resulted in frequent drought conditions.

Agricultural Impact: The recurring drought has severely affected agriculture in Bundelkhand. Crops fail due to water scarcity, leading to loss of livelihoods for farmers. The lack of irrigation facilities and inadequate access to water exacerbate the agricultural crisis in the region.

Water Scarcity: The persistent drought has led to acute water scarcity in Bundelkhand. Many water sources, such as rivers, reservoirs, and groundwater aquifers, have depleted or dried up. This has made it challenging for people to access safe and sufficient drinking water.

Livelihood Challenges: The drought has caused significant challenges for the local population's livelihoods. Agriculture, which is the primary source of income for many, has been severely impacted. Migration in search of alternative employment has become common, leading to social and economic disruptions in the region.

Government Interventions: The governments of Uttar Pradesh and Madhya Pradesh have implemented various relief and rehabilitation measures to address the Bundelkhand drought. These include providing financial assistance to affected farmers, distributing food and water, implementing water conservation projects, and supporting alternative livelihood options.

Need for Long-Term Solutions: The ongoing drought in Bundelkhand highlights the need for long-term solutions to address water scarcity and improve the region's resilience to drought conditions. These solutions include promoting sustainable farming practices, implementing efficient irrigation systems, constructing rainwater harvesting structures, and improving water management and conservation techniques.

Awareness and Support: Various organizations, NGOs, and civil society groups have been working to raise awareness about the Bundelkhand drought and provide support to affected communities. These initiatives focus on water management, livelihood diversification, and sustainable development in the region.

The Bundelkhand drought is an ongoing challenge that requires concerted efforts from the government, local communities, and stakeholders to address water scarcity, agricultural distress, and the overall well-being of the affected population. Long-term solutions and sustained support are crucial to mitigate the impact of drought and promote sustainable development in the region.

Case Study 15

The Pathankot airbase attack refers to a terrorist attack that took place on January 2, 2016, at the Pathankot Air Force Station, a major military installation located in Pathankot, Punjab, India. The attack was carried out by a group of heavily armed militants believed to be affiliated with the Pakistan-based militant group Jaish-e-Mohammed. Here are some key details about the Pathankot airbase attack:

Incident: The attack began when a group of militants infiltrated the airbase, which houses fighter jets and other military assets. The militants breached the perimeter fence and engaged in a gun battle with security forces.

Casualties: The attack resulted in the loss of several lives. Indian security forces, including soldiers and commandos, were involved in the operation to neutralize the militants. Several security personnel, including soldiers and airmen, were killed or injured during the counter-operation.

Duration and Resolution: The operation to neutralize the militants lasted for several days. It involved intense gun battles, hostage situations, and coordinated efforts by Indian security forces to secure the airbase. The operation concluded with the successful elimination of the militants.

Investigation and Intelligence Cooperation: Following the attack, investigations were conducted to identify the perpetrators and determine the extent of their network. India requested cooperation from Pakistan, as the attack was believed to have been planned and executed from across the border. Intelligence sharing and cooperation between the two countries were crucial in the investigation.

Impact on Bilateral Relations: The attack had significant implications for India-Pakistan relations. It led to a temporary suspension of the bilateral dialogue process, as India demanded concrete action against the militant group responsible for the attack.

Counterterrorism Measures: The Pathankot attack highlighted the need for enhanced counterterrorism measures and improved security protocols at sensitive military installations. It prompted a reassessment of security arrangements, intelligence coordination, and strategic responses to such terrorist threats.

Continued Security Concerns: The attack on the Pathankot airbase served as a reminder of the ongoing security challenges faced by India and the need for sustained efforts to counter terrorism and ensure the safety of critical installations.

The Pathankot airbase attack in 2016 was a tragic incident that resulted in the loss of lives and raised concerns about cross-border terrorism. It underscored the importance of robust security measures, intelligence cooperation, and efforts to address the root causes of terrorism for maintaining peace and stability in the region.

Case Study 16

The Pulwama attack refers to a devastating terrorist attack that took place on February 14, 2019, in the Pulwama district of Jammu and Kashmir, India. The attack targeted a convoy of vehicles carrying Indian security personnel. Here are some key details about the Pulwama attack:

Attack: A suicide bomber drove a vehicle laden with explosives and rammed it into one of the vehicles in the convoy. The convoy consisted of vehicles carrying Central Reserve Police Force (CRPF) personnel.

Casualties: The attack resulted in the loss of 40 CRPF personnel who were killed in the blast. Many others were injured in the explosion.

Responsibility: The Pakistan-based militant group Jaish-e-Mohammed claimed responsibility for the attack. The attacker was identified as a local militant belonging to Jaish-e-Mohammed.

Condemnation and International Response: The Pulwama attack received widespread condemnation from the international community, including neighboring countries. There were calls for holding the perpetrators accountable and taking measures to combat terrorism.

India's Response: India condemned the attack and vowed to take appropriate action against the perpetrators. In response, India carried out airstrikes on terrorist camps in Balakot, Pakistan, targeting Jaish-e-Mohammed training facilities.

Diplomatic Relations: The attack strained India-Pakistan relations, with India blaming Pakistan for providing support and a safe haven to terrorist organizations. There were diplomatic tensions between the two countries, including the suspension of cross-border trade and travel.

Security Measures: The Pulwama attack prompted a reassessment of security arrangements and counterterrorism strategies in the region. Measures were taken to strengthen intelligence gathering, border security, and preventive actions against militant groups.

Commemoration and Remembrance: The Pulwama attack is remembered as a tragic incident in the history of India's fight against terrorism. Commemorative events and tributes are held annually to honor the sacrifice of the CRPF personnel and to show solidarity with their families.

The Pulwama attack in 2019 was a horrific act of terrorism that resulted in the loss of many lives. It had far-reaching implications for the region's security dynamics and India-Pakistan relations. The attack highlighted the ongoing challenge of cross-border terrorism and the importance of collective efforts to combat such threats.

Case Study 17

The COVID-19 pandemic is an ongoing global health crisis caused by the novel coronavirus SARS-CoV-2. It was first identified in December 2019 in Wuhan, Hubei province, China, and has since spread to virtually every country in the world. Here are some key details about the COVID-19 pandemic:

Spread and Impact: The virus spread rapidly through human-to-human transmission, primarily through respiratory droplets when an infected person coughs, sneezes, or talks. It led to a wide range of symptoms, from mild respiratory illness to severe respiratory distress, pneumonia, and multi-organ failure.

Global Health Emergency: The World Health Organization (WHO) declared the COVID-19 outbreak a Public Health Emergency of International Concern on January 30, 2020, and subsequently declared it a pandemic on March 11, 2020. This recognition highlighted the urgent need for global coordination and response to contain the virus.

Public Health Measures: To mitigate the spread of the virus, public health measures were implemented worldwide, including widespread testing, contact tracing, quarantine and isolation protocols, social distancing, travel restrictions, and the promotion of hand hygiene and mask-wearing.

Healthcare System Strain: The pandemic placed an enormous strain on healthcare systems globally. The high number of COVID-19 cases overwhelmed hospitals, leading to shortages of critical medical supplies, beds, and healthcare personnel. Efforts were made to increase healthcare capacity and develop effective treatment strategies.

Economic Impact: The pandemic had severe economic repercussions, with numerous industries and businesses affected. Lockdowns, travel restrictions, and disruptions in global supply chains led to job losses, business closures, and economic recessions. Governments implemented fiscal stimulus packages to support individuals and businesses.

Vaccines and Vaccination Efforts: Multiple vaccines were developed and authorized for emergency use to combat COVID-19. Vaccination campaigns have been launched worldwide to administer vaccines to as many people as possible, prioritizing high-risk groups and healthcare workers.

Social and Psychological Impact: The pandemic had a profound impact on people's mental health and well-being. Isolation, fear, grief, and the socio-economic consequences of the pandemic increased stress levels and exacerbated mental health issues. Efforts were made to provide mental health support and promote resilience.

Global Collaboration: The pandemic highlighted the importance of global collaboration in addressing public health crises. Scientists, researchers, and healthcare professionals worldwide collaborated to share knowledge, exchange best practices, and develop treatments and vaccines.

The COVID-19 pandemic remains an ongoing global challenge, with efforts continuing to mitigate its impact, control the spread of the virus, and ensure widespread vaccination. The pandemic highlighted the significance of robust public health systems, effective crisis management, and international cooperation in addressing future health emergencies.

Case Study 18

The swine flu outbreak in 2009, also known as the H1N1 influenza pandemic, was a global health crisis caused by a new strain of the influenza A virus. Here are some key details about the swine flu outbreak:

Emergence: The outbreak was first detected in April 2009 when a new strain of the H1N1 influenza virus was identified. The virus was a result of genetic reassortment, combining genetic elements from pig, bird, and human influenza viruses.

Global Spread: The swine flu quickly spread from its origin in Mexico to other parts of the world. The World Health Organization (WHO) declared it a pandemic on June 11, 2009, indicating that the virus had spread across multiple continents.

Symptoms and Transmission: Swine flu had similar symptoms to seasonal influenza, including fever, cough, sore throat, body aches, and fatigue. It was primarily transmitted through respiratory droplets, similar to other influenza viruses.

Global Response: Governments and health organizations worldwide implemented measures to control the spread of the virus. These included surveillance, diagnosis, treatment, contact tracing, and public health campaigns to raise awareness about preventive measures such as hand hygiene and mask-wearing.

Impact: The swine flu pandemic had a significant impact on public health, healthcare systems, and the global economy. It caused illness and fatalities, particularly among vulnerable populations such as young children, pregnant women, and individuals with pre-existing health conditions.

Vaccination Campaigns: Vaccines were developed and distributed worldwide to mitigate the impact of the swine flu pandemic. Vaccination campaigns targeted priority groups, including healthcare workers, high-risk individuals, and later, the general population.

Evolution and Subsequent Seasons: Following the initial outbreak, the H1N1 influenza virus became one of the seasonal flu strains, circulating alongside other influenza viruses. It is now included in annual flu vaccines to provide protection against H1N1.

Lessons Learned: The swine flu outbreak highlighted the importance of early detection, rapid response, and global collaboration in addressing pandemic threats. It emphasized the need for robust surveillance systems, vaccine development capabilities, and public health preparedness.

Since 2009, swine flu has continued to circulate as a seasonal influenza virus. Ongoing surveillance, vaccination efforts, and public health measures help to mitigate its impact and reduce the severity of outbreaks.

Case Study 19

The Uttarakhand forest fires in 2016 refer to a series of devastating wildfires that occurred in the state of Uttarakhand, India. The fires spread across various forested areas, causing extensive damage to the environment, wildlife, and human settlements. Here are some key details about the Uttarakhand forest fires in 2016:

Timing and Extent: The forest fires occurred during the summer months of April and May in 2016. The fires affected large parts of the state, including forested regions in the districts of Nainital, Almora, Pauri Garhwal, Chamoli, and Tehri Garhwal.

Causes: The primary cause of the forest fires was a combination of factors, including high temperatures, low humidity, dry vegetation, and human activities such as encroachment, illegal land clearing, and negligence in fire management practices.

Environmental Impact: The forest fires had severe consequences for the environment. The fires destroyed vast stretches of forest cover, leading to soil erosion, loss of biodiversity, and disruption of ecosystems. The release of carbon dioxide due to the burning of vegetation also contributed to air pollution and climate change concerns.

Wildlife Impact: The fires had a detrimental effect on wildlife habitats. Many animals, including rare and endangered species, lost their natural habitats and were displaced or killed. The loss of vegetation also impacted the availability of food and shelter for wildlife.

Human Impact: The forest fires posed significant challenges to local communities and human settlements in the affected areas. They led to the destruction of houses, agricultural fields, and infrastructure. The fires also had adverse effects on the health of people, as smoke and pollutants from the fires caused respiratory problems and other health issues.

Firefighting Efforts: The Uttarakhand forest fires prompted a massive firefighting operation. Various agencies, including the Indian Air Force, National Disaster Response Force (NDRF), State Forest Department, and local volunteers, were involved in extinguishing the fires and preventing their further spread.

Rehabilitation and Reforestation: Following the fires, efforts were made to rehabilitate the affected areas and restore the damaged ecosystems. Reforestation initiatives, including planting of saplings and promoting community involvement, were undertaken to rejuvenate the forests.

Prevention and Preparedness: The forest fires highlighted the need for improved fire management practices, including early detection, swift response, and better coordination between government agencies and local communities. Awareness campaigns and training programs were conducted to educate people about fire safety and prevention measures.

The Uttarakhand forest fires in 2016 were a significant environmental and human crisis, resulting in the loss of valuable forest cover, wildlife habitats, and human settlements. The incident emphasized the importance of sustainable forest management, community involvement, and proactive measures to prevent and manage forest fires in the future.

Case Study 20

In 2019, Himachal Pradesh, a state in northern India, did experience forest fires that had significant implications for the region. Here are some key details about the forest fires in Himachal Pradesh in 2019:

Timing and Extent: The forest fires occurred primarily during the summer months of May and June in 2019. Several districts in Himachal Pradesh were affected by the fires, including Kullu, Shimla, Mandi, Chamba, and Kinnaur.

Causes: The primary causes of the forest fires were a combination of factors, including dry weather conditions, high temperatures, low humidity levels, and human activities such as negligent behavior, discarded cigarette butts, and open burning practices.

Environmental Impact: The forest fires had severe consequences for the environment. They resulted in the destruction of vast stretches of forest cover, leading to soil erosion, loss of biodiversity, and disruption of ecosystems. The release of carbon dioxide due to the burning of vegetation also contributed to air pollution and climate change concerns.

Wildlife Impact: The fires had a detrimental effect on wildlife habitats in Himachal Pradesh. Many animals, including species endemic to the region, lost their natural habitats and were displaced or killed. The loss of vegetation also impacted the availability of food and shelter for wildlife.

Human Impact: The forest fires posed risks to human settlements and communities in the affected areas. They led to the destruction of houses, agricultural fields, and infrastructure. The fires also had adverse effects on human health, as smoke and pollutants from the fires caused respiratory problems and other health issues.

Firefighting Efforts: The forest fires triggered a massive firefighting operation in Himachal Pradesh. Various agencies, including the State Forest Department, Indian Air Force, National Disaster Response Force (NDRF), and local volunteers, were involved in extinguishing the fires and preventing their further spread. Helicopters and other aerial firefighting resources were deployed to control the fires.

Rehabilitation and Reforestation: Following the fires, efforts were made to rehabilitate the affected areas and restore the damaged ecosystems. Reforestation initiatives, including planting of saplings and promoting community involvement, were undertaken to rejuvenate the forests.

Prevention and Preparedness: The forest fires highlighted the importance of improved fire management practices, including early detection, swift response, and better coordination between government agencies and local communities. Awareness campaigns and training programs were conducted to educate people about fire safety and prevention measures.

The forest fires in Himachal Pradesh in 2019 had significant environmental, wildlife, and human impacts. They underscored the need for sustainable forest management, community involvement, and proactive measures to prevent and manage forest fires in the future.

Impact of Natural Disaster in India

Natural disasters in India can have significant impacts on various aspects of life. Here are some of the common impacts of natural disasters in India:

Loss of Human Life: Natural disasters can result in the loss of human lives, causing immense grief and trauma for families and communities.

Displacement and Migration: Disasters such as floods, cyclones, and earthquakes can lead to the displacement of people from their homes, forcing them to seek temporary shelter in relief camps or migrate to safer areas.

Infrastructure Damage: Natural disasters can cause severe damage to infrastructure, including houses, buildings, roads, bridges, and communication networks. This disruption can hinder rescue and relief operations and impede normal functioning of essential services.

Economic Loss: Disasters can have a significant economic impact, leading to the destruction of crops, livestock, and agricultural land. Industries may suffer losses due to damage to factories, interruption of supply chains, and decreased productivity.

Environmental Degradation: Natural disasters can cause environmental degradation, such as deforestation, soil erosion, water pollution, and habitat destruction. This can have long-term ecological consequences, affecting biodiversity and ecosystem services.

Public Health Issues: Disasters can exacerbate public health challenges. Contamination of water sources, inadequate sanitation facilities, and disrupted healthcare services can lead to the spread of waterborne diseases, vector-borne diseases, and other health risks.

Social and Psychological Impact: Natural disasters can have a profound impact on the social fabric of communities. Disrupted livelihoods, loss of homes, and psychological trauma can lead to social and psychological distress among the affected population.

Impact on Education: Disasters can disrupt the education system, causing the closure of schools and universities. Students may face challenges in accessing education, resulting in learning gaps and long-term educational setbacks.

Strain on Resources: Natural disasters place a strain on resources and emergency response mechanisms. Governments, humanitarian organizations, and communities need to allocate resources for rescue, relief, and rebuilding efforts.

Climate Change Vulnerability: India's vulnerability to natural disasters is further amplified by the impacts of climate change. Rising temperatures, changing rainfall patterns, and sea-level rise can increase the frequency and intensity of certain disasters.

Efforts are made by the Indian government, NGOs, and international organizations to strengthen disaster management, improve early warning systems, and enhance disaster preparedness to mitigate the impacts of natural disasters and promote resilience in affected areas.

Natural Disaster

Natural disasters refer to destructive events that are primarily caused by natural processes or forces. These events can have severe consequences on human life, property, and the environment. Here are some common types of natural disasters:

Earthquakes: Sudden shaking or trembling of the ground caused by tectonic plate movements beneath the Earth's surface.

Hurricanes and Cyclones: Intense tropical storms characterized by strong winds, heavy rainfall, and storm surges. They form over warm ocean waters.

Floods: Overflow of water onto normally dry land due to heavy rainfall, dam failure, or rapid snowmelt.

Tornadoes: Violently rotating columns of air that extend from a thunderstorm to the ground, causing significant damage in their path.

Wildfires: Uncontrolled fires that rapidly spread across vegetation and forested areas, often fueled by dry conditions and strong winds.

Tsunamis: Large ocean waves generated by underwater earthquakes, volcanic eruptions, or landslides that can cause devastating coastal damage when they reach the shore.

Volcanic Eruptions: Outpouring of molten lava, ash, and gases from a volcano, which can lead to ashfall, pyroclastic flows, and lahars.

Landslides: Mass movement of rocks, debris, and soil down a slope, often triggered by heavy rainfall, earthquakes, or human activities.

Droughts: Prolonged periods of abnormally low rainfall, leading to water scarcity, crop failure, and ecosystem disruption.

Avalanches: Rapid downslope movement of snow, ice, and rock, often triggered by snowpack instability or other factors.

It's important to note that while natural disasters can be devastating, preparedness, early warning systems, and effective response measures can help mitigate their impact and save lives.

Natural Disaster Mapping and Management

Natural disaster mapping and management involve the use of geospatial data, technology, and strategies to understand, predict, mitigate, and respond to natural disasters. These activities aim to minimize the impact of disasters on human lives, infrastructure, and the environment. Here are key aspects of natural disaster mapping and management:

Hazard Mapping: Hazard mapping involves identifying areas prone to specific types of natural disasters, such as floods, earthquakes, hurricanes, wildfires, or landslides. Geospatial data, including historical records, topography, climate patterns, and satellite imagery, are used to create maps that indicate the likelihood and intensity of different hazards in specific locations.

Vulnerability Assessment: Vulnerability assessment involves evaluating the susceptibility of a region, infrastructure, and population to natural disasters. Factors such as population density, land use, building codes, infrastructure resilience, and socioeconomic conditions are analyzed. This information helps identify high-risk areas and vulnerable communities that require targeted mitigation efforts.

Risk Assessment: Risk assessment combines hazard mapping and vulnerability assessment to estimate the potential impact of natural disasters. It involves quantifying the potential loss of life, damage to infrastructure, economic impact, and environmental consequences. Risk assessment helps prioritize resources, plan evacuation strategies, and allocate funds for disaster preparedness and response.

Early Warning Systems: Early warning systems utilize real-time monitoring, sensor networks, and modeling to detect and forecast natural disasters. These systems provide timely alerts and notifications to communities at risk, enabling them to take appropriate actions and evacuate if necessary. Early warning systems are critical for mitigating the impact of fast-onset disasters like hurricanes, tsunamis, and severe storms.

Emergency Response Planning: Disaster management involves developing comprehensive emergency response plans that outline roles, responsibilities, and coordination mechanisms among various stakeholders. These plans define protocols for evacuation, search and rescue operations, medical assistance, infrastructure protection, and post-disaster recovery. They also address communication strategies to ensure effective dissemination of information during emergencies.

Remote Sensing and GIS Technologies: Remote sensing and geographic information systems (GIS) play a significant role in natural disaster mapping and management. Satellite imagery, aerial photography, and remote sensing data are used to monitor changes in the environment, track natural hazards, and assess damage after disasters. GIS technologies help analyze and integrate geospatial data to facilitate decision-making and resource allocation.

Community Engagement and Education: Community participation and awareness are essential for effective natural disaster management. Engaging with communities through public education campaigns, training programs, and drills enhances their preparedness and response capabilities. Promoting a culture of resilience and providing information on evacuation routes, emergency shelters, and communication channels are crucial for saving lives during disasters.

Post-Disaster Recovery and Reconstruction: After a natural disaster, mapping and management efforts extend to post-disaster recovery and reconstruction. This involves assessing the damage, identifying areas for reconstruction, developing land-use plans that account for future hazards, and implementing measures to enhance resilience. It also involves supporting affected communities in rebuilding their lives, infrastructure, and livelihoods.

Natural disaster mapping and management are dynamic processes that involve continuous monitoring, assessment, and adaptation. By leveraging technology, data, and community engagement, effective disaster management strategies can save lives, protect assets, and build resilient communities in the face of natural hazards.

Important Questions

1. What is the main natural disaster that commonly occur and require mapping and management?
2. How can mapping technologies and satellite imagery help in predicting and monitoring natural disasters?
3. What are some effec5tives strategies for disaster management and response after a natural disaster occurs?
4. How can Geographic Information System be utilized in natural disaster mapping and management?
5. What are the key challenges in accurately mapping and managing naturel disasters?
6. How do early warning systems contribute to natural disaster management and preparedness?
7. What role can remote sensing play in assessing the impact of natural disaster and guiding response efforts?
8. How can social media and crowdsourcing be leveraged for real-time information gathering during natural disasters?
9. What are some examples of successful natural disaster management initiatives from around the world?
10. How can governments and communities collaborate to enhance natural disaster mapping and management efforts?

Index